THE SCIENCE OF CULTURE
AND THE PHENOMENOLOGY OF STYLES

The Science of Culture and the Phenomenology of Styles

RENATO BARILLI

Translated by Corrado Federici

McGill-Queen's University Press
Montreal & Kingston • London • Ithaca

Authorised translation from the Italian language edition
published by Bononia University Press
ISBN 978-0-7735-4099-6 (cloth)
ISBN 978-0-7735-4104-7 (paper)

Legal deposit fourth quarter 2012
Bibliothèque nationale du Québec

Printed in Canada on acid-free paper that is 100% ancient forest free (100% post-consumer recycled), processed chlorine free.

The translation of this book has been funded by SEPS – Segretariato Europeo per le Pubblicazioni Scientifiche (Via Val d'Aposa 7 – 40123 Bologna, Italy, www.seps.it, seps@seps.it).

McGill-Queen's University Press acknowledges the support of the Canada Council for the Arts for our publishing program. We also acknowledge the financial support of the Government of Canada through the Canada Book Fund for our publishing activities.

Library and Archives Canada Cataloguing in Publication

Barilli, Renato
[Scienza della cultura e fenomenologia degli stili. English]
The science of culture and the phenomenology of styles / Renato Barilli ; translated by Corrado Federici.

Translation of: Scienza della cultura e fenomenologia degli stili.
Includes bibliographical references and index.
ISBN 978-0-7735-4099-6 (bound).–
ISBN 978-0-7735-4104-7 (pbk.)

1. Art and society. 2. Art and history. I. Federici, Corrado, 1946– II. Title. II. Title: Scienza della cultura e fenomenologia degli stili. English.

N72.S6B3413 2012 701'.03 C2012-905474-7

This book was typeset by True to Type in 10.5/13 Sabon

Contents

Introduction 3

1 The Concept of Culture 9

2 The Stratum of "High" or Symbolic Culture 24

3 The Internal Articulations of Cultural Systems 46

4 The Birth and Evolution of "Modern" Space 60

5 The Birth and Evolution of the Contemporary Age 77

6 Heinrich Wölfflin and the Phenomenology of Styles 136

Postscript 153

Annotated Bibliography 157

Index 175

THE SCIENCE OF CULTURE
AND THE PHENOMENOLOGY OF STYLES

Introduction

This work, which appears in English translation some thirty years after it was originally published, should be readily accepted by the North American public since virtually all the works on which my thinking is based were produced in that part of the world, beginning with a classic text from which I borrowed the first part of my title, *The Science of Culture*, by Leslie A. White (1949). I read the Italian version published in 1969 and it has become a cornerstone for my own theoretical investigations. In the first edition of the present volume, I took the more radical step of adopting the term culturology, whereas in the second, I opted in favour of a neutral phrase to avoid upsetting the general readers and the students to whom the book was addressed. White provided an excellent overview of the decades of research carried out in the field of cultural anthropology, research that reappears in condensed form in another classic work, *Culture: A Critical Review of Concepts and Definitions*, co-authored by Clyde Kluckhohn and Alfred L. Kroeber (1952). All these writings define the concept of culture on the basis of the human capacity to produce implements or prostheses to extend the capacity of the human body – in other words, external implements designed to enable humans to intervene on the environment. From this premise arises the important concept of cultural materialism, on which another famous cultural anthropologist, Marvin Harris, has theorized elegantly in an essay titled *Cultural Materialism* (1979). In brief, I have completely revised my initial assumptions in the sense that, to find the essence of culture, we need to look outside ourselves and to invert St Augustine's dictum: no longer *Noli foras ire, in te ipsum redi, in interiore homine habitat veritas* (Do not desire to go outside. Return in yourself. The truth dwells inside your consciousness), but rather *Ex te ipso exi, in exteriore homine habitat veritas* (Go outside yourself. The truth dwells outside your consciousness). All this wisdom culminated in the thoughts of Canadian Marshall McLuhan, whose work I encountered when his most complete work and his most fully-developed ideas appeared in Italian translation. This was *Understanding Media* (1964), of which I became the apologist in Italy and described specific implications derived from that work for the domains of art and literature, to which my activity as scholar and professor was dedicated. The phrase "The medium is the message," which has often been criticized for being banal and superficial, seemed, on the contrary, enlightening to me, to the point where it resolved in the best way possible the epistemological issue of our age by building on Kant's legacy and his idea of synthetic a priori judgment: we shape external data and experience through the media we adopt. As a result, we have a synonymous term in the sense that the media of a given epoch are simply the technology used in that particular time and place.

These anthropologists who served as my guides were very careful not to lapse into the nineteenth-century determinism found in both Positivism and Marxism, but they did acknowledge that these two great currents of thought captured at least one aspect of the solution by underscoring the fact that we have to start from the ground up, from material reality. However, this premise did not ensure the progress of human beings because, in addition to being involved with the here and now and intervening on the environment with the implements provided to us by the technology we use, we have the capacity to question and speculate on our future. This occurs thanks to the adoption of a particular instrument, the symbol. It too is based on material reality – otherwise, our general presupposition would be contradicted. In fact, the symbol comprises phenomenal data, which are the visual elements in art or written literature, and sounds in music and drama. A symbol is something that is not physically present but rather located in a hypothetical or virtual dimension. All this leads us to refute the hegemony that semiotics, i.e. a general science based on the primacy of the sign, has claimed over the entire western world from the 1960s onwards. The dense cluster of anthropological theories to which I am indebted is clear in this regard: the sign is not the exclusive product of human beings. On the contrary, we share this instrument with animals. Therefore, semiotic studies are blameworthy or defective when they presume to relegate the symbol to the level of performing minor or secondary tasks and fail to acknowledge the vital role that the symbol actually plays.

At this point, I also have to recognize the role that the best philosophical European tradition played in my own intellectual development and in the composition of the pages that follow. If I am inclined to interpret the symbol as the expression of a uniquely human capability, I cannot ignore the best analysis on this topic provided by German philosopher Ernst Cassirer in his *Philosophy of Symbolic Forms* (1923), including the impressive expansion of this theory on the part of Susanne Langer, an American philosopher of art, born of German parents. She deserves special credit for having explained that symbols are divided into two large families: presentational and referential symbols. Presentational symbols pertain to the arts since they are based on a strict linkage between the material and the virtual strata; whereas with referential symbols, the relationship between the material and the virtual components is conventional and, as such, they are particularly suited to the sciences. This division, which is also well known to European scholars who study these matters, is the basis of my own *Course on Aesthetics*, which I had the good fortune of seeing translated and published by the University of Minnesota Press in 1993. I would like to add that the large family of referential symbols is divided into two subcategories: those that are analytical and used in the physical and mathematical sciences, and those that have a more physical quality since they impact on the senses, as occurs in rhetoric. My own work in this particular area, namely *Rhetoric*, was also translated and published by the University of Minnesota Press in 1999. So I am proud to claim that my presence in North America is thus established from this moment onwards through a trilogy of works that form an interdependent "system."

Returning to the European context, Cassirer himself points the way to establishing a direct link with criticism and art history. Consider the figure of Erwin Panofsky and his *Perspective as Symbolic Form*, published in 1927, where he adapts the symbol with consummate skill to applications in the artistic field. Panofsky was the last exponent of a noble European tradition, found particularly in Germany, which includes such brilliant stars as Alois Riegl and especially Heinrich Wölfflin, with his famous dialectical pairs. This branch of European philosophy and art history had a serious limitation, however, in that it completely neglected the connection between the "high" and noble stratum of the symbolic forms and the "low" stratum of material practices. Panofsky and his adherents could be accused of omitting, or indeed setting aside, all the problems revolving around the relationship between the material and the virtual or the symbolic. But they did so for the very good reason that they did not know how to solve the problem. We close our eyes when we find ourselves faced with a dark and threatening reality we cannot control.

From here, I return to the answers provided by the cultural anthropologists mentioned above, as well as to the impact their ideas had on my own thinking. Such authors as White and Harris firmly established the fact that culturology must proceed on two tracks: the material-technological and the virtual-symbolic. But they were also aware of the difficulties associated with connecting these two strata. At this point, it is useful for me to go back across the Atlantic to find the best solution to this problem in the theories of French sociologist Lucien Goldmann, who, while especially interested in the history of narrative, did not exclude the possibility of transferring and adapting his tenets to broader areas of inquiry. Goldmann wanted to have his cake and eat it too. In other words, he wanted to account for both the enormous weight of the material practices of a given epoch and the singular dignity of the symbolic forms. To this end, he asserted the existence of a homology between the two processes, or the identity of the respective practices, which can be illustrated if we extract from a given discipline a structure, a pattern, or modality reduced to schematic terms. We may even speak of the possibility of congruence, as occurs in geometry, which is the superimposition of the schemas derived from each individual sector for the purpose of determining their properties. The adoption of an invaluable concept like homology still seems to me the best possible solution to this nagging problem.

I revert once more to the teaching of White where he acknowledges the priority of common discoveries over the exploits of individual protagonists in a given cultural period. The times, circumstances, and modalities adopted are more powerful than individual talent, or at least it can be argued that talent can facilitate or accelerate the attainment of certain objectives, but not create them from nothing. We are profoundly influenced by the times in which we live and from this realization stems my personal preference for the "generational criterion," which is also implied in White's theories and those of North American culturologists: When the time is ripe, we inevitably come across similar solutions, which should be more correctly called homologies, in different places. It is possible for many individual talents to arrive simultaneously at a particular discovery or invention, even though they are not in contact with one another

These issues should be well known to readers in North America, as I state in the opening paragraph. They may be surprised or puzzled, however, by the periodization of historical cycles, which is to say, the succession of different cultures. In Europe, at least in Italy, our textbooks used to be constructed on the basis of chronologies or timetables that I believe have no equivalent in North America. They have been nearly forgotten in Europe due to the general policy of deregulation of our educational system and the vagueness of the terms adopted, which renders them less incisive than they ought to be. We are accustomed to speaking of a modern age that begins in 1492 with the death of Lorenzo the Magnificent and with Christopher Columbus's discovery of America and that ends in 1789 with the start of the French Revolution; it is followed by the contemporary age, an age that has not drawn to a close. However, the terms "modern" and "contemporary" are synonyms; in effect, they indicate two phases of modernity. Furthermore, in terms of the very concept of cultural materialism explored here, it would be better to refer to technological events. Here, I once more accept fully what McLuhan stated indirectly, although he was never concerned with questions of chronology – the "modern" period corresponds to the Gutenberg era and should begin in 1450 with the invention of the movable type printing press, to which McLuhan himself correlates the homological event that occurred a few years earlier at the symbolic level, namely, the theorization of perspective by Leon Battista Alberti in 1435 on the basis of the inverted pyramid. To indicate the start of the next period, we need to identify a world dominated by the technological exploitation of electromagnetic phenomena. The first discoveries and inventions in this field emerge toward the end of the 1700s in the work of Luigi Galvani (animals as condensers of electric energy) and Alessandro Volta (the battery and the first electromagnetic generator), among others. These events remained at the symbolic level for the time being since they were scientific theories lacking practical technological applications. This eloquently explains the reason for which the "old" modern age continued to dominate during the 1800s – as did machines propelled by thermal energy – together with the persistence of symbolic forms still founded on the perspectival pyramid, subsequently confirmed by the camera itself.

The textbooks are correct, then, both in the chronological relationship between these two periods (perhaps with some slight adjustment of the dates) and in the consideration of these periods as distinct from each other and not interchangeable. But what remains is the semantic emptiness of the two terms with which we try to characterize the two epochs. I personally tend to prefer getting rid of the term "contemporary" and substituting it with the term "postmodern," which contains within it an indicator of succession and, therefore, of difference. However, in order to proceed along this path, the term postmodern should be pushed back many years to have it begin in the late 1700s – the time of the first experiments in electrology (the application of electromagnetism and electrotechnics). These experiments in turn could be seen as homologous to certain features in the art of Johan Heinrich Füssli (Henry Fuseli) and William Blake, namely, forms completely removed from the perspectival pyramid that undulate in the foreground the way electromagnetic waves do. Some may raise the objection that the term postmodern is already used to designate a spectrum of phenomena that, to tell the truth, is rather confused and incoherent, beginning during the 1960s. If one were to accept my argument, the postmodern would no longer be the initial phase but a late, autumnal phase of something that started much earlier. Today, early in the twenty-first century, as I state in the postscript to the present English translation, we are at the point of maximum expansion of the electronic world and, consequently, at the point of maximum extension of the postmodern, which by now impacts the entire planet, making the global village theorized by McLuhan relevant as never before. But the term "village" has a precise meaning and it takes us back to the idea of limited space, closeness to home, and as such it deprives globalization of its potential to flatten differences. Today's artistic media, such as photography, video, installations, and new modalities of writing, which are widespread, enable the world to recover its roots. We are, therefore, on the verge of attaining a felicitous synthesis of the two opposing terms, a phenomenon that can be described with that widely diffused term "glocalism."

I

The Concept of Culture

FROM STYLUS TO STYLE

Each term in the title of this book undoubtedly requires some clarification but, contrary to what one may believe, it is not the most complicated terms, such as phenomenology, that require an explanation. We will see, instead, that words like phenomenology surrender their secrets rather easily, whereas the more common and humble terms, culture and style, which are part of everyday speech, appear to be among the more controversial and difficult to pin down. This does not occur only in the case of these two words, but rather constitutes a general principle, which is that common or widely-used words are loaded with multiple, ambiguous meanings and, therefore, greater effort is required to interpret them.

Naturally, whenever we want to shed light on the meaning of a particular word, it is useful to consult a good dictionary. In the case of *stile* [style] this elementary process of consultation produces a result that could be repeated in other cases of this type; that is to say, meanings are created initially at the material level, in concrete living conditions or situations, where there is intervention on the environment. This corresponds generally to the literal meaning of a word, which is accompanied almost always by a second or derived sense resulting from a transfer of meaning to broader and more general fields increasingly removed from the concrete situation giving rise to that meaning. In fact, we speak of a level of translated or transposed meanings (these terms derive from two different Latin root words associated with the idea of bringing or carrying, with the added suggestion of movement). We can also refer to the corresponding Greek root word and speak of "metaphorical" meanings. There is also the possibility of calling these meanings figurative, but we would be operating within the same set of concepts since these figures are fictional and replace real, present elements thereby allowing us to imagine real elements that are absent. Therefore, let us acknowledge that almost every word has this dual nature, this transfer of meaning from an original, material one to a broader, more abstract one.

Returning to the word style, *stile* in Italian, etymological dictionaries tell us that the literal and material ancestor of this word is *stylus*, which is an implement of practical life, in other words, a practical instrument. Specifically, a stylus is a small metallic shaft, pointed at one end, used to etch clay tablets covered with a layer of wax. During the Roman era, these tablets were the principal support for writing. The tablets produced for common occurrences were different from those produced for occasions that were meant to be immortalized, in which case, capital letters were carved on the hard surface of stone.

Today, people interested in archaeology and the history of writing are particularly attentive to that remote word *stylus* (*stylo* in Vulgar Latin). The Italian word has undergone a minor change, almost a softening; as a result we have *stile*, which is "style." The word has experienced a typical transfer from the material to the abstract plane, a process that can still be understood by going back to the original term. The employment of the stylus, an exquisitely practical activity in itself, from the very beginning made it possible to introduce expressive variants into the act of writing. Each scribe had his own way of writing or making the letters while respecting the conventions of writing. In other words, each copyist produced what could be considered to be his own personalized handwriting: an individual style. Here, we need to be careful not to affirm the individuality of the writer too hastily. Scribes active in the same time period, and sometimes in the same place or workshop, shared the teachings of a common source and had to respect writing norms. Right from the start, though, we need to avoid setting in opposition the individual and the collective because the two can exist very well in sequence, which is to say one after the other or in a series of superimposed strata, like the leaves of an artichoke. Without a doubt, there was a graphic style common to all the scribes working in the same scriptorium or school, but this does not mean that each could not develop individual or eccentric characteristics. Even today, when we write longhand (no longer using a stylus but a ball-point pen), with every word we write we unconsciously blend features shared by a whole group or sector of society to which we belong, with features that pertain to our own individual personality. In short, from the initial material activity involving the stylus, we see the emergence of activities that are increasingly general and refined, thus allowing individuals and communities to make expressive choices. From the style of writing in the graphic context, we pass to style in verbal expression (the types of words and sentence structures we select, for example – do we prefer coordination and short sentences or subordination and clauses that are layered and arranged like waterfalls?). From the verbal, we can shift to other fields, provided it is possible to operate in them respecting pre-established social or institutional norms while exercising individual or group preferences. As a consequence, there can be a style of dress, movement, dining, entertainment, and general behaviour. In the circumstances created by the community in which we live, the metaphorical process expands exponentially and comes to constitute a continent that is increasingly removed from the material context where it originated.

For all these reasons, we can see the essentially pluralistic nature of style. It would be more appropriate, therefore, to speak of styles since each group, generation, social class, and individual tends to exercise his or her options even if this means only small but tangible deviations from what the group or the person next to him or her is doing. Furthermore, this plurality occurs not only in synchrony but in diachrony as well, in as much as styles already differentiated in a given historical period are destined to undergo change as they travel into subsequent historical epochs. We are dealing with unstable and changing forms to which the term "phenomena" should be applied; by this I mean that things appear to us here and now and are different from the way they appeared there and then, and they in turn were in the process of changing.

Thus, we have the opportunity to examine in greater detail the term "phenomenology," whose apparent complexity but essential clarity at the denotative level I have already mentioned. In particular, the suffix "logia" is not a mystery since it is commonly used to indicate that one intends to apply a systematic discourse or "logos" to the object identified by the lexeme preceding the suffix; so we have the study of "phenomena," phenomenality being the variability of these phenomena across time. In order to establish a phenomenology, we must determine that a given field is characterized by forms that appear and disappear, forms that have a limited duration, in other words entities that resolve themselves in appearance, so that the phenomenal dimension is more important than the ontic (to use the Greek term). On the other hand, the logical component inevitably involved in every scientific enterprise is aimed at identifying both stability and precariousness; therefore, wherever it is applied, phenomenology aims to bring a level of stability to the precarious elements. Here, it is not my intention to deny the infinite number of styles, but rather to ask if there are recurring characteristics within this incessant process of mutation. It is true that styles change, but is it possible to identify criteria of investigation or frames of reference despite the changes? This is the general task of every phenomenology and a phenomenology of styles is no exception, because it is within the purview of every undertaking of this kind.[1]

THE MACRO-CONTINENT OF CULTURE

Broadening our study, we finally arrive at the macro-continent of culture, which is typically where human beings establish their institutions of every type, constructing, changing, criticizing, and undermining them from within, even as they create new ones. In a sense, we can say that styles emerge entirely within culture itself, which therefore ultimately functions as its container.

Naturally, it is better to do away with the more learned term culturology, replacing it with the more common expression "science of culture," as indeed I do several times throughout the present book. Culturology, or more conventionally the science of culture, refers to the intention of subjecting to a systematic and scientific analysis an object that in itself is difficult and ambiguous, i.e., culture, precisely because of its simplicity and widespread use. Even the person in the street accepts the term and makes it part of his or her daily vocabulary. That same person, however, might find himself or herself rather embarrassed if asked to provide a clear, precise definition of the term. The same word also presents serious difficulties for people who con-

1 In addition to designating in general the systematic study of all phenomena, the term phenomenology refers to a particular philosophical movement founded by Edmund Husserl, in which the investigation of "appearances" for the purpose of uncovering their properties is of fundamental importance. See Annotated Bibliography, Chapter 4.

sider themselves sufficiently cultured or well educated. Even in this case, it is wise to return to the dictionary, particularly to establish the etymology; in doing so, we will find a clear confirmation of the primacy of the material meaning, which proves to be significant in the present investigation of the definition of culture.

Etymology tells us that, in the Latin world, culture was typically a material-physical intervention on the land, and even today we cannot refer to such interventions without using words derived from that Latin root. I am referring to the cultivation of fields or *agriculture*, and it is appropriate to say that one of the primary forms of human culture has been agriculture. We should also be grateful to the Italian language for the fact that it introduced only a minor deviation to the Latin, changing slightly the root lexeme *cult* to *colt* (as in *coltura*) to designate practical, material interventions on the environment, leaving the more archaic *cult* to signify metaphorical or figurative processes that have to do with working at the level of ideas or cognition and are removed from the immediate material situations that gave rise to them.

Or is this distinction something negative destined to be added to the many prejudices that, throughout the centuries, have distanced the material from the abstract or figurative meaning of culture? Certainly, one of the first steps in modern culturology is to perform an open-minded, rigorous, and in-depth analysis of the object that is culture. In order to do this, we must begin by returning to the material level. Culture must rediscover the pride of its origins, which lie in physical interventions on the rough and difficult conditions of the environment.[2]

From this idea comes one of the first conditions for defining the concept of culture in a rigorous and succinct manner. It is peculiar to humankind (it does not make sense to speak of the culture of animals) precisely because culture is linked to the typically human capac-

2 Among the many studies on the subject, it is worth mentioning what has become a classic, Clyde Kluckhohn and Alfred L. Kroeber's *Culture: A Critical Review of Concepts and Definitions* (New York: Vintage Books, 1952). Another seminal work is Marvin Harris' *The Rise of Anthropological Theory* with the telling sub-title *A History of Theories of Culture* (1968; repr., Walnut Creek, CA: Altamira Press, 2010). The more recent work by the same author, *Cultural Materialism* (1979; repr., Walnut Creek, CA: Altamira Press, 2001) is also very important.

ity of knowing how to work, which in turn is made possible by the adoption of inorganic implements, which are called inorganic in as much as they are not an organic part of the human body. The human being is an animal that extends its limbs with objects taken from the external world. To refer to these extrinsic implements, we can use the term prosthesis, which is somewhat depressing, given its current circulation, relating as it does to orthopaedic processes whereby limbs are replaced; yet the term is an effective one precisely because it designates a typical human trait: the ability to enhance the natural physical apparatus with artificial parts added to that body, much like orthopaedic prostheses, which can be removed after use and certainly cannot be integrated with the natural organism; nor can they be transmitted from parent to child through the genetic code (but, it is obviously possible to transmit the potential for re-acquiring them once again through the cultural code).

From their origins, humans have found it useful to intervene in their environment to extract food, defend themselves from adverse climate conditions, attack their neighbours, and protect themselves against aggressors by extending the natural body with external objects, which are in turn capable of giving rise to a systematic series of other interventions in the environment. This sequence corresponds to work, or the establishment of cycles of intelligent, coherent, and specifically designed applications, for which the Greek language gives us the root word *techne* and the Latin language gives us the equivalent *ars*. In short, an enormous array of technical advances begins to take shape. These coincide with the presuppositions of culture itself, at its material level. Humans possess this capacity to adopt tools, non-organic prostheses, and to use them to develop the sphere of technology. The great anthropological ages always mentioned in our handbooks – Palaeolithic, Neolithic, Eneolithic, and Bronze – are simply vast technological cycles of material culture, which were fundamental for the emergence and flourishing of humanity. Each of these instrumental ages corresponds to specific expressive styles; these are the choices made by potters, armourers, and other artisans determined to forge their own tools and products.

Seen in this broad historical perspective, the claim that the Latin language designates culture with a term relating to the process of working the land turns out to be partial, somewhat misleading, and not as satisfactory as one might think. In fact, culturology and its immediate predecessor, cultural anthropology, tell us that the first implements and the first phases of human material culture were not devoted to working the fields, but rather to such nomadic activities as hunting and fishing; therefore, humans produced prostheses or technical apparatuses adapted to perform these tasks. Agriculture comes later and does not necessarily imply an advantage for the populations affected by this system.[3] This is where we find prototypical, sedentary societies in which hierarchies or class distinctions are cemented. The risk factors certainly diminish as does the unpredictability of the surroundings but only to the extent that social roles and the distinction between dominator and dominated, hegemonic and subaltern classes, is too rigid, to the disadvantage of the latter. This would be the model of the industrial-mechanical society that dominates the nineteenth century and much of the twentieth.

The axiomatic statement that there is culture where there is the capacity to employ non-organic implements seems indispensable to a definition of culture, but it is not sufficient. This is because the immediate objection arises that some species of animals are capable of employing objects of these types. Ants accumulate dirt in order to construct their anthills; bees laboriously organize the honeycomb inside their hives; dogs can be trained to retrieve a stick thrown by their owners; and of course monkeys, given their opposable thumbs and toes, are agile and able to do things in ways that humans cannot. How can we distinguish ourselves from these capable competitors, the animals? How can we make culture an exclusively human prerogative? Alternatively, would it be better to expand the boundaries of culture and accept the fact that even animals can perform technology? And what can we say about bird nests and spider webs? There is also a much greater difficulty; how would we avoid that these prostheses or objects invented by humans do not remain static and unchangeable across time?

Perhaps a definition of culture would suffice if we added a second equally necessary condition encompassing this human ability not only to create prostheses, but also to change them through time. One day, people use a stone as it presents itself to them in nature. The next day, they discover that the stone melts when it is subjected to high

3 This reassessment of nomadic cultures in relation to sedentary ones is attributable to another enlightening work by Marvin Harris, *Cannibals and Kings* (1977; repr., New York: Vintage Books, 1991).

temperatures; then it can be fused with other materials extracted from the ground and, once solidified, the composite becomes especially resistant and solid, and so forth. In this way, we open up the glorious path of the history of technology.

THE MATERIAL STRATUM AND THE IDEAL STRATUM

The transition from one type of prosthesis to another implies the opening up of a dimension of intelligence, or at least the ability to observe, plan ahead, and visualize scenarios different from those unfolding in the moment. Humans can suspend the blow they are about to strike with their primitive wooden hammer and its rudimentary handle and wonder about other possibilities, first conceptualizing then realizing these possibilities through experimentation and study. In this way, we pass from the stratum of material, strictly technological-instrumental culture to a second stratum, which is derived or figurative, in other words, the phenomenon recorded in dictionaries, which are reliable repositories of a diachronic process unfolding across centuries and millennia. Thus, we arrive at the ideal, noetic dimension of "thought," in which many theorists are quick to identify the specificity of culture and, therefore, the defining core of the entire phenomenon.

At this point, the advantages of the first of the two postulates need to be recalled. The stratum of intelligent expression, which is undoubtedly the basis of ideation, cannot be cut off entirely from its instrumental nature, i.e., the use of non-organic or external prostheses, on which I have found it useful to base the entire continent of culture. The original technical practice used by humans to express themselves – graffiti and stone drawings – represents a common external repository that relieves organic memory of essential tasks and avoids the degradation of acquired knowledge, which is inevitable if it is consigned only to organic memory. From the moment they are born, animals have to begin all over again to accumulate experience since they do not have a pre-existing databank. They can rely only on the information already deposited in their genetic code; very long periods of time are required for the code to absorb additional data – a timespan that goes well beyond a phenomenal concept of history. This explains why the behaviour of animals that depend on a similar genetic code does undergo noticeable changes, even if these are distributed over thousands and perhaps millions of years. In an observable period of history, there are no evident changes; therefore, we may come to the conclusion that these changes belong to the sphere of culture, rather than to the sphere of nature. Humans are the only animals able to take advantage of a non-organic memory delegated to external material bodies that last longer than a human lifetime. In this way, children and other descendants discover them, acquire them rapidly, integrate, enrich, challenge, and revise them according to new parameters.

Culture, therefore, is inevitably articulated in two levels, two strata or functional spheres. One is to be considered strictly material and concerns the sphere of technical practices involving the development of implements; the other concerns itself with the sphere of "flights of the imagination," where humans think of future possibilities, imagine scenarios different from those currently in effect and ask what would happen if they used implements or different practices with respect to those being used at the time. This level corresponds to the ideal, conceptual culture, or whatever we choose to call it. To a certain extent, this sphere also relies on material supports, which underpin broader areas of activity, and these can be called "ideal" or symbolic.

Given the substantial homogeneity of the two environments and levels, which are part of the same continent of culture, the suspicion may arise that there may be no point in making a distinction between the two. Instead, I argue that a distinction is useful, in addition to being obvious. Many readily acknowledge the difference between a hammer and a philosophical, political, or artistic idea. Of course, we must be careful to avoid committing two opposing errors. One is to believe that the hammer is strictly nature or matter, overlooking the fact that it is a materialized idea. The other error is to consider ideas as spaces of pure contemplation, separate from the practical life. We must also appreciate the specific characteristics of each. In order to function, the implements used to intervene in the environment for the purpose of extracting its resources and transforming them, transporting them, etc., invariably have materiality and inertia. As a result, these implements appear as "other than the self" to workers, i.e., those who have to put them to use. Implements can often resemble despotic rulers who impose strict rules and demand labour from workers without granting them any autonomy. In short, tools acquire the characteristics of a second nature or a "language" that the individual speaker cannot change or modify because it is stronger than the speaker and it exercises a power that can shape him or her. This is almost the myth of Frankenstein, the monster that escapes the control of its creator, who sees the monster turn on him with its inexorable mass of muscles, like a brute force.

The ideal stratum, instead, represents a situation where there is no direct physical, material interaction with nature and where the function of the implements is translated into symbols and subjected to study and reflection – for the purpose of improving, phasing out, or replacing them with more efficient implements. From this emerges the space of what is commonly referred to as culture, but culture is also susceptible to excessive idealism if understood as a luxury of the spirit as opposed to a laboratory where the new is designed and a correlation is established among the implements employed in various sectors.

Having established the fact that these two spheres exist, which is useful to do for a number of reasons, the problem of their reciprocal relationships arises (an ideal classical problem where culture undertakes to investigate itself, thereby giving rise to the science of culture or culturology). There are horizontal relationships among the various sectors of the same stratum (for example, within the sphere of culture as it is currently understood, which is to say the ideal meaning of the term, relationships would involve epistemology, psychology, ethics, aesthetics, etc.). These interconnections have been widely investigated and satisfactory results have been produced, although these are clearly related to the different ways in which these relationships have been described and studied. Almost no one today doubts the usefulness or the legitimacy of investigating these types of relationships. Much more doubtful and polemical, instead, is research on vertical types of relationships, those between the implements adopted on the "high" stratum and those adopted on the "low" stratum of material technology. From this arises the fear that we may lapse into a banal determinism that subordinates one stratum to the other – in this case the "high" stratum being subordinated to the "low" – making the latter simply a mirror image of the former. Conversely, if we underestimate the inertial mass of the "low" stratum, we run the opposite risk of idealism, which is to say material implements understood as docile projections of the will. In the present study, I strive to avoid these two extremes, keeping in mind the fact that material technology is still culture, the product of ideas or human planning. In order to avoid lapsing into idealism, however, we must remember that technology enjoys autonomy, which in turn can become more powerful than the person who institutes that autonomy, and it can bring to bear upon that person a force that has unpredictable consequences.

The most useful and effective concept to characterize these types of vertical relationships between the two strata of culture was proposed by Lucien Goldmann[4] who employed the term "homology." This concept ensures some key elements:

a) Equal dignity of both strata, between which a relationship that we could call homological is established. Thus, we would remove the risk of determinism, which we would face if we were to contend that the ideal "mirrors" the material. This aspect of the question was especially interesting to Goldmann and to a large sector of contemporary "culture" (high culture) since that culture was in large part Marxist, which means that it tended to think in terms of the relationship between the economic base and the ideological superstructure. In particular, Goldmann saw the need to avoid subordinating the latter to the former, which occurs in the theories of his acknowledged teacher, Georg Lukács.
b) Equality, which is neither identicalness nor external resemblance. The superstructure (the ideal stratum) does not have to produce an image that conforms to the one occurring on the "low" stratum, or at least identicalness shifts from external features to function, assuring the right of each sphere to have its own, different experiences that are reciprocally incongruous and sector-specific. Here we are well beyond the notion of "representation" or a passive mimetic relationship. In order to perceive a homology, we need to refer to deep structures or functional criteria on the relevant strata and sectors. It is necessary to extract models from the data clearly; only after such a phase of extrapolation is it possible to determine if there is the possibility of superimposing one pattern onto the other, thereby demonstrating a reciprocal congruency beyond the respective particularities and surface features.

4 For a more in-depth study of Lucien Goldmann's theories, especially his notion of homology, see *Towards a Sociology of the Novel* (1964, Trans. Alan Sheridan; repr., New York: Tavistock Publications, 1975). See also my critical commentary, on which the methodology used here is based, in the chapter dedicated to him in *L'azione e l'estasi* [Action and ecstasy] (Milan: Feltrinelli, 1967; repr., Turin: Testo & Immagine, 1999).

THE CONCEPT OF TECHNOMORPHISM

How does this line of thinking affect the visual arts? One of the primary tasks of a culturological study is to search for homological relationships between works or processes of a given epoch and the material technology that dominates or emerges in that same epoch. Obviously, it is also possible to imagine a scenario where periods are dominated by a single technology as well as periods in which there is friction and conflict between old and new technologies. Such an inquiry must avoid the false presupposition that there can be homologies only in cases where the work of art transcribes, imitates, or represents the visible features of the technology of a given period. Instead, these visible aspects may simply be temporary, non-essential features, the product of imitation of other epochs and other strategies of rationalization. It is, therefore, entirely up to the visual "operators" of a certain historical or cultural epoch to find visual, plastic, and spatial forms that make tangible and visible the logic inherent in the techniques used at the material level. In short, it is up to the visual arts to produce technomorphism, a term to be taken literally, meaning to give plastic, spatial form to technological procedures used in the same historical period. In other words, since my analysis focuses on critical, theoretical observations, technomorphism becomes the working hypothesis for those who are conducting research and formulating critical interpretations.

Helpful in this regard is Erwin Panofsky's famous notion of "symbolic forms" (proposed in the context of Renaissance perspective), which is expressive of the concerns identified in the preceding paragraphs. In defining symbolism, Panofsky[5] wanted to avoid running the risk of naturalism and mimeticism. He underscored that the Renaissance concept of space was a cultural construct and the product of mathematics, astronomy, philosophy, and art combined, and that the form of the construct produced by painters and architects was not an image in a representational sense, but a symbol, which is to say an invention, an autonomous externalization, though one capable of making those ideas visible.

From this point of view, however, Panofsky was hampered by the fact that he limited his correlations to different sectors of "high" cul-

5 Erwin Panofsky's famous essay is from 1927, *Perspective as Symbolic Form* (Cambridge, MA: MIT Press, 1991). See also the Annotated Bibliography and Chapter 4.

ture and ignored completely those of "low" culture, that is, the stratum of material technology. This does not mean that his valuable concept of "symbolic forms" cannot be adopted by those who feel the need to tackle this aspect of the relationship between the various sectors of the same culture. The technomorphic manoeuvre, which I have identified as the primary task of the visual arts, therefore, translates into the task of inventing "symbolic forms" that transform the working mode and the tools of the base technology into images, visual compositions, and plastic, spatial products. The unification of models does not negate autonomy because it is the special and exclusive task of the visual arts to find the visual and plastic symbols of operational schemas that, in other sectors, can be assigned to the anaesthetic (in the literal sense, without suggesting any connection to *aesthesis*) and intangibility of ideas or the crudeness of implements that, in their external form, can mask or even contradict their own internal logic.

THE THEORIES OF MARSHALL MCLUHAN

At this point, however, the question arises: Is it possible to go a step further in an attempt to establish a homological correlation between the "high" and "low" strata of culture? Or, is this a risky undertaking with no guarantee of producing scientific results? There are those who have tried to take this additional step, achieving results that are convincing, even if they are formulated as condensed thoughts, bursts of insight, aphorisms, and paradoxes, as opposed to organic statements. I am referring in particular to Marshall McLuhan, whose theories I have embraced totally and enthusiastically over the past fifty years or so, in an effort to make his ideas more accessible and systematic.[6] For example (and to return to the unquestioned achievements of Panofsky), the material stratum of technology that can be correlated to the symbolic form of Renaissance perspective is precisely what McLuhan called the Gutenberg Galaxy, the movable type printing press. This technology introduces functional criteria based on the homogeneity of the ele-

6 The two works by Marshall McLuhan to which I make constant reference in these pages are *The Gutenberg Galaxy* (Toronto: University of Toronto Press, 1962) and *Understanding Media* (1964; repr. Cambridge, MA: MIT Press, 1994). For a more thorough introduction to McLuhan, see the chapter I devote to him in *Tra presenza e assenza* [Between presence and absence] (Milan: Bompiani, 1981).

ments used and their repetitiveness, as well as on the possibility of arranging these elements serially in lines and columns. All of this is homological to the conceptual grid (of "high" culture) based on Cartesian coordinates, which are instruments used to measure or "analyze" infinite and homogenous space in each of the three dimensions. In addition, for it to be read, the printed message was thought to compel readers to place themselves at a certain distance and to assume a point of view, projecting onto that surface a "beam of linear optical rays." From here, we jump to the homology between the act of reading and the inverted pyramid or the *camera obscura*, which are the pivotal concepts on which Renaissance perspective rests.

In this instance, we also find positive features that necessarily accompany every application of the homological paradigm:

a) The printing press does not determine Renaissance perspective nor is it its cause. McLuhan was the first to point out that it would be completely groundless to suggest this, given that the "operators" (at the ideal stratum) who invented Renaissance perspective, including architects and painters, were almost always active prior to the actual invention of Gutenberg's printing press, normally dated to the mid-1400s. There is, therefore, no record of earlier events at the "low" stratum, as far as homologies on the "mental" side are concerned. By contrast, there is co-operation between equals, having equal dignity, with chronological lags that may vary since the "operators" in one sector can anticipate the achievements of those in another sector without making their convergence on a common objective impossible.
b) This convergence or teleology is the working hypothesis I am proposing here as an *a posteriori* interpreter in a privileged position of observation (like someone who observes from the top of a hill or from a helicopter and sees that vehicles on different roads are about to converge at an intersection, without the drivers necessarily being aware of this). We do not have to presume that "operators" in their respective sectors are aware of what is happening in adjacent sectors. Basically, one form of creativity impacts on every point of the cultural sphere and the influence is, therefore, measurable to varying degrees in each sector. Consider the classic phenomenon of the communicating vessels, where the liquid reaches the same level irrespective of the diameter of the vessels.

c) Continuing with this analogy, the vessel that pertains to the material technology of an epoch is larger, has a greater diameter, and the various aspects of a culture are visible in it at the macroscopic level; similarly a flooding river makes hydrometers used to measure the overflow unnecessary. Beyond the analogy, it is worth repeating that a decisive role is played by "ideas" when they are represented in material form. In this case, they attain a weight or urgency, whereas they appear to be precarious and fragile when they are first announced at the pure level of calculation and laboratory experiment.

It can be argued that Gutenberg's technology (which is appropriate for the information sector) is compatible with that of the machine powered by the thermal energy produced by vapour or combustion, used in the production and transportation sectors. Renaissance perspective, which is the result of the felicitous fusion of the Cartesian coordinate system and the concept of the inverted pyramid, is the symbolic form of both. Perspective serves mechanomorphism satisfactorily, that is to say, the form or symbolic visualization of the entire mechanical era, characterized at the material level by the development of Gutenberg's printing press and industry based on machines powered by heat energy. These are events that have corresponding homologies at the "high" level of culture in the period beginning with Renaissance perspective and ending with Impressionism, which coincides with the invention of the optical and photochemical processes that give birth to photography.

Naturally, McLuhan does not stop here. His next step is to describe the emergence of a very different technological cycle than the mechanical one in which Gutenberg operated. This new cycle represents a quantum leap in quality and I am referring to the great electronic and electrotechnical era. For developments in this area and the problems that pertain to it, see Chapter 5.

2

The Stratum of "High" or Symbolic Culture

SYMBOLIC SYSTEMS

Having discussed the material stratum of culture, it is now time for me to turn to the ideal or symbolic stratum, which is to say, the fields that manipulate ideas rather than material objects and correspond to the various disciplines and sciences that house the knowledge of a given historical period. According to many, this stratum is culture in the most current usage of the word.

I do not wish to devote too much time and space to defining the word "idea" since there are entire libraries on this topic and since philosophy, the philosophy of knowledge or epistemology in particular, has concerned itself with the subject for centuries. No one can doubt the existence of these general modes of action that embrace entire groups, families, or object types, while recognizing that such a vast area implies the loss of concreteness. Ideas cannot be touched or seen, contrary to the etymology of the word (from the Greek *idein*, to see). This is not an insignificant etymology, given that sight is the most dematerialized of the senses because it lends itself to such a range of applications and, therefore, is analogous to the special role played by ideas. From this stems the variable fortune of ideas depending on the cultural contexts in which the ideas appear. If the contexts privilege the dimension of tangible facts or objects, ideas appear to be shadows or unreliable, imperfect *simulacra*, thereby representing a minus rather than a plus; at any rate, they are derived or secondary products. If these contexts, instead, disparage manual or practical work, ideas become noble entities worthy of free human beings, especially if these individuals are surrounded by throngs of slaves and servants to whom they can delegate the task of satisfying their basic natural needs. One of the objectives of culturology is precisely to see how such variations are configured in the relationships between the material and the ideal strata.

Today, a balanced judgment tends to prevail, giving each stratum its due, which is to say, recognizing the dignity of general forms or ideas while not depriving them of their material base. To this end, the etymology of the verb linked to the act of seeing is useful in that it also refers to physiological behaviour (the contraction of the intraocular lens, the constriction of the pupil, the appropriate disposition of the body, etc.). There is also mental or internal seeing; for example, thinking of the idea of horse. While distinct from physiological behaviour, internal vision appears to be accompanied by bodily movements, even though these are nearly atrophied. This is because, even in such a context, humans have always been quick to benefit from their natural ability to acquire or invent non-organic implements as we have seen occur at the level of material culture. What distinguishes the human condition from that of animals is not the capacity to extrapolate (even a dog, for example, can conjure up the idea of food from a series of material circumstances presented to it), but the capacity to entrust the idea to a small amount of matter that is light, practical, and useful, and can be preserved. In short, I am referring to the very well-known ability to "signify" or construct signs (elements that are concise and easy to manipulate) that stand for things or circumstances that are more cumbersome and inconvenient. This sphere of signs in turn introduces an enormous set of problems and issues that I do not intend to discuss at this point.[1] In fact, I will leave aside the notion of sign since it is still too generic (we usually employ it to refer to material things rather than ideas: smoke is a sign of fire) and reflect on the more specific and subtle concept of symbol, which is understood as something physical (graphic or auditory) that stands for an idea or a general concept (the tricolour flag that stands for the country of Italy and H_2O that is a symbol for the chemical structure of the water molecule).

1 Mention of Umberto Eco's *A Theory of Semiotics* (Bloomington: Indiana University Press, 1976) is obligatory. On the issue of the connection between semiotics and the culturological perspective, see my contribution to *Estetica e società tecnologica* [Aesthetics and technological society] (Bologna: Il Mulino, 1976), the subtitle of which is rather eloquent: *Difficoltà di un approccio semiotico alla culturologia* [The problem of a semiotic approach to culturology].

The capacity to produce signs can be found everywhere, in things as well as animals, whereas the capacity to produce symbols is peculiar to humans. (Ernst Cassirer calls the human being *animal symbolicum*[2] or symbol-making animal). This takes us back to what I said in the previous chapter since the use of symbols is simply the extension of the human capacity to produce non-organic implements to the domain of ideas. The ideas that animals have are governed by their bodies, their sensory system, and are entrusted to an organic memory that has such limitations as precariousness, perishability, and the impossibility of transmitting information externally. Animals inherit only the knowledge that has been incorporated into the genetic code, and so they can be reduced to the level of instinct and non-culture or what the elderly are able to learn in a short period of time. Humans, instead, have at their disposal an external, objective system of memorization that provides a greater capacity for storing various kinds of information. Here, we have two notions that are usually invoked to define human culture or culture per se: memorization and symbolization. At any rate, we are dealing with two notions that go together and perhaps constitute only one notion, if we are talking about an external memory entrusted to material symbolic systems. However, we should not rush to identify these systems with the alphabetical systems of writing. We know that material culture in such an environment has much to tell us. Graphic memorization systems do not always and in all cultures prevail over oral or auditory systems.

Finally, these two convergent features explain the other trait that distinguishes human culture from animal behaviour, a trait I have already emphasized. This is the capacity to produce new things, to advance new ideas, which in turn are materialized into new tools or, conversely, to deploy new technologies that can only lead to an altered symbolic projection. How can we explain such a capacity if not in

2 See Ernst Cassirer *Philosophy of Symbolic Forms* (New Haven & London: Yale University Press, 1965, 4 vols). Also important are the studies devoted to him by Susanne Katherina Langer, *Philosophy in a New Key* (1942; repr., Cambridge: Harvard University Press, 1979) and *Feeling and Form* (1953; repr., New York: Prentice Hall, 1977). See also note 11. Culturologist Leslie White in his book *The Science of Culture: A Study of Man and Civilization* (1949; repr., New York: Farrar, Strauss and Giroux, 1969), also prefers the notion of symbol to define the "high" level of culture. See also Sandro Briosi, *Il simbolo e il segno* [The symbol and the sign] (Modena: Macchi, 1993).

terms of the advantages presented by the storage of received notions, which is made possible by symbolic memory? Humans have at their disposal the accumulated knowledge of their predecessors in a convenient and easily accessible format; therefore, they can devote their energies to enriching that inheritance and to going beyond it rather than memorizing it with their organic memory. In such a scenario, previous experiences are preserved and consolidated, whereas in the case of animals, there is an enormous waste of information.

By now, the reader will have noticed the constant care I take not to create an unbridgeable gap between the material and the ideal stratum, thereby avoiding the reappearance of those pointless and dreary debates that, in a variety of cultural contexts, have shaken the entire edifice of knowledge, namely, the debate on matter versus spirit, naturalism versus idealism, empiricism versus rationalism, and so forth. These are interactive phenomena within a single macro-system, which is culture as a whole, so we must ask ourselves if it is worthwhile drawing a distinction between these two strata, as though we were performing the impossible surgical procedure of separating the two sides of the same reality. In fact, when we concentrate on material culture, that is, on the technology of an epoch, are we not really speaking of a sort of applied epistemology in effect? In other words, are we not cleverly making a substitution, replacing practical implements with ideas? It is, therefore, inevitable that a homology results between ideas materialized in practical instruments and ideas seen in their role as proxies, which is to say, philosophy and mathematics books. For example, is the Gutenberg technology (according to McLuhan the pillar of the modern age) not perhaps described here in such a way as to be Cartesian epistemology *ante litteram* (which is to say, analytic geometry or a logical-mathematical construction of space) of the *res extensa*? Or in other words, a precursor to Cartesian notions of apprehending materiality? At the beginning of the contemporary or postmodern age, we have certain technological innovations, such as Pacinotti's ring armature and the transoceanic telegraph cables, or perhaps the equations of Maxwell, who mapped out the electromagnetic field. Why should we operate on two strata when, in reality, we are dealing with only one: that of the ideal or symbolic systems, seen in a pure sphere at one point and in the applied sphere at another.

However, it is possible to continue to assert the usefulness of a distinction, provided that the result is not an inflexible distinction. We are dealing with a dynamic relationship between two strata (or a dialectical relationship, as we would have spoken of it in the past) involving circularity or feedback, where one stratum nourishes the other in an infinite series of actions and reactions. Ideas and innovations alter the way we work, produce, and communicate, but these activities influence, stimulate, and condition one another.

ARTS, SCIENCES, AND FACULTIES

I now turn to the main purpose of the present study, a close examination of the stratum of knowledge, its internal subdivisions, and its relationship to the material stratum, with the aim of identifying homologies between ideal and material culture. To reduce even further the potential gap between the two, I would point out that an initial articulation in the field of knowledge is the distinction made among the various arts, if the term is understood in the light of the Latin root *ars*, which is very close to the Greek *techne*. Moreover, here we find ourselves nearly in the presence of a "coup de théatre" that could immediately re-open the dilemma examined in the preceding paragraphs. Either the artistic (technical) systems are kept out of the culture of the ideal stratum and nudged into the material-technological stratum or they remain within the material sphere, but in this case they erase all possible distinctions and ensure the existence of a continuum, of the gradual and quantitative transition from the arts to the sciences or from technology to pure research. The answer is never absolute or "ab-solved" of the conditions existing in the cultural context. For example, in ancient and medieval times, there was undoubtedly a continuum that ensured the cultural status of all the arts, but tended to exclude them from the so-called "superior" culture of the content sciences and the discourse sciences, considered to be the only ones worthy of the free individual. Etymologically, both *ars* and *techne* entailed manual activities carried out with intelligence, ability, and qualitative excellence, and were aimed at producing material objects. The fact that these objects were then used to produce practical things or things that were in turn symbolic (for example, frescos and sculptures) represented a difference, of course, but it was part of a condition that remained generally that of the manual labourer or artisan; therefore, practitioners were relegated to an inferior social rank, very often considered to be unworthy of the "free" individual and instead left to slaves or freed slaves. This situation applied to the artisanal production of visual objects as well as such acoustical objects as music, and even theatrical performances, such as those of mimes and actors in tragedies or comedies – activities considered unworthy of the full citizen or unworthy of feudal lords, clergymen, and members of the emerging middle class during the Middle Ages. The only exception was the art of letters because it concerned responsibilities primarily of an ethical and political nature; it was, therefore, advantageous to be highly specialized or competent in letters. Furthermore, at the oral level, but to a large extent even in writing, the profession of letters certainly does not get one's hands dirty; rather, it is considered to be an ideal activity that must be distinguished from manual, servile practices.

The arts as we understand them today, that is to say the fine arts, have undergone various phases of ennoblement and progressive detachment from the servile, minor, or mechanical arts (from craftsmanship) and have been inserted fully into the ideal sphere. The first phase occurred during the historical process that goes by the name of the Renaissance and marks an increase in the idealist aims of certain arts (painting, sculpture, and architecture), which had the task of realizing high, qualitative values (beauty, rigour, harmony, and equilibrium). However, this does not yet lead to a change in the social status of the artist, who remains tied to his or her artisanal origin, even if that origin is largely forgotten or set aside. It is a fact that, although they were praised, Raphael and Michelangelo were not considered worthy to sit at the table of popes and princes, while "operators" at the verbal level, the humanists who revived the noble arts of poetry and rhetoric, were.

An even more important phase, indeed a decisive one, occurs in conjunction with the first Industrial Revolution, marked by machines powered by heat energy (the end of the 1700s and the first half of the 1800s), a revolution that, according to the bold theories of McLuhan, constitutes a sort of secular arm of the Gutenberg era, its complete realization at the level of productivity. As we know, machines capable of mass-producing stereotypical merchandise replace craftsmanship to a large extent, putting an end to that useful intermediate stage which, until that point in time, assured a correlation between quantity and quality, between material and ideal culture. Consequently, we witness the collapse of the social status of the artisan, who undergoes the well-known process of "proletarianization" when required to work in factories, where we find working conditions that often make the inhuman conditions for slaves in ancient history pale in comparison.

(Under the guise of freedom or liberty, the worker is exploited by the "steam barons" in a much more systematic and thorough measure). At that point, the artist (the practitioner of the noble arts of painting, music, or drama) is pushed upwards, definitively removed from a servile condition (that of worker or proletarian), and finally inscribed into the "superior" level but with no economic or professional guarantees. This is also the case because, if in the artisanal technological system there was harmony between the low and the high strata, now the fine arts seem to be exercised in opposition to the stereotype plate of the printing press. The artist works against the system of production and becomes an agent of protest, one who dreams of non-conformist ideals and better times. This is a profession that he or she would have to exercise at his or her own peril, even if it is fascinating and beautiful precisely for this reason; in fact, at the same time as the first Industrial Revolution culminates, the artistic avant-gardes (those now referred to as "historical" because they date back to the last century) come into being, functioning like a system of antibodies to this new status quo.

It is no accident (at least, if the theories developed here hold) that the clean split between mechanical technology and standardized industrial production on the one hand, and artistic images of different (avant-garde) systems on the other, occurs in the 1860s when some exceptionally sharp minds are able to detect the advent of an alternative technology founded, as we know, on electronics and its applications.

From that moment on, the opposite process begins, no longer a violent clash between the vile mechanical arts and the libertarian experiments of the artists, but rather a reconciliation and osmosis, precisely because of the possibility that the same material technology can acquire an appreciable degree of quality (aesthetic in the broad sense). In addition, we should not overlook an important intermediate, mechanomorphic phenomenon, still within the modern period. While there was the belief that mechanical technology would remain an invariant for many decades, certain artistic styles, such as Cubism in painting or the Modern Movement in architecture, tried to reconcile mechanomorphism and aesthetic canons so as to heal the gaping space opened at the beginning of the process, when it seemed that the quantitative, stereotypical, repetitive logic of the machine would engage in a no-holds barred battle with the aspirations of the fine arts. Today it appears that we once again have a continuum that stretches from artistic production or technology, according to the etymology of the word, to a superior, experimental, and innovative sphere with "high" aesthetic responsibilities inherited from the traditional exigencies of the fine arts, even if the concept of beauty has now been eclipsed.

FORMAL SCIENCES AND CONTENT SCIENCES

Within the system of ideal or symbolic culture, the arts have always played an intermediate role as a potential link with the material stratum, like a Trojan horse for both. This is not true for those disciplines that have always had special reflective responsibilities within the realm of ideas, almost the task of administering them, providing instruments that can be widely applied and are not tied to specific subjects. These are formal disciplines and they are distinct from other disciplines, which are tied more directly to their object of study and take the name of content disciplines or sciences, in the current meaning of the term. Precisely because of their broad applicability, the first are also called "faculties." Due to the fact that the prevailing symbolism in the various historical cultures is of the verbal-discursive type, for the most part, these are considered disciplines, faculties, or forms of discourse. Furthermore, since the Greek equivalent of the Latin root word for "verbal" or discursive is *logos*, we have the logical forms. The other meaning of *logos*, i.e., "road" or "way," enters into this equation as well. Logistics is also the field that concerns itself with practical matters – for example, studying the ways in which roads can be provided for the transportation of goods and people. Additionally, we have the other widely used Greek term *method* (the way along which we pass). Thus, we can identify the block of disciplines or formal faculties that are logistical or methodological in nature and tend to cut across the content disciplines.

In the western tradition, Aristotle was the first great codifier of the various logistical pathways. Let us avoid even the slightest suspicion that the history of culture contains definitive or unsurpassed bodies of knowledge; such an idea is far removed from the present form of reasoning. Aristotelian logic, therefore, seems quite distant from our way of thinking, because it ignores the great technological changes that have occurred in the 2,000 years that separate us from Aristotle. In terms of abstract projection and typological models that can be imagined, however, the resources are not infinite and the human imagination is not inexhaustible. This means that certain dyads and triads, on which the Aristotelian philosophy is constructed, have the virtue of being very economical and, therefore, are still serviceable today, but only as frameworks for which different content has to be provided. In any case, what is important here is to establish typologies of evolution and change, rather than inflexible classifications that are fixed once and for all.

The fact remains that the three forms of discourse identified by Aristotle have had a very long life and we can still make use of these nearly timeless categories because of their suitability and economy. So, to be brief, we continue to be confronted with three kinds of discourse: analytical, dialectical, and rhetorical.

Analytical reason applies to sciences that study objective content external to man, for example, physical or natural content, and it presupposes that it is possible to formulate postulates, "true" axioms from which we can infer equally true consequences using rigorous procedures. Dialectical reason is directed primarily at the principles or postulates of the sciences and investigates their truth claims, revealing that the dialogue within a community of scholars in many cases both lays the foundations and dismantles them. There are, therefore, no fixed truths, only degrees of probability, which is to say the capacity to bring about a high degree of consensus among the interlocutors through certain arguments and debate. Up to this point, however, analytics and dialectics share the property of being forms appropriate for rigorous scholars cloaked in the silence of their laboratories. However, if the search for consensus occurs in a forum, before an assembly of peers in whose hands rest the criteria for establishing whether a person or action is beautiful or ugly, useful or useless, just or unjust, we find ourselves in the presence of rhetorical reason.[3]

An important aspect of the search for homological relationships is the investigation of the ways in which material culture or technology influences these three sister forms of logic or discourse, determining their fortune, their rise and fall, their hegemonic tendencies as well as their attempts to expand to the point of absorbing the others or risk-

3 I refer the reader to my book *Retorica* [Rhetoric] (Bologna: Fausto Lupetti, 2011), previously in *Corso di retorica* [Course on rhetoric] (Milan: Mondadori, 1995), in which I examine the fortunes of rhetoric relative to those of analytics and dialectics. Giulio Preti's, *Retorica e logica* [Rhetoric and logic] (Turin: Einaudi, 1968) is of fundamental importance in this regard.

ing elimination. For instance, being characterized by a low level of technological innovation and few changes in the physical-mathematical sciences on the one hand, and on the other by an active public affairs sector, in the broadest sense of the word (politics, economy, military strategy) on the part of an oligarchic social group, the classical Roman period witnesses the unprecedented growth of rhetoric, which became the *ratio* capable of absorbing the other two forms, considered as secondary. (Cicero does not hesitate to attack the physical sciences with the weapon of statistical probability, denying that they are capable of attaining truth).[4]

The late Middle Ages present a radically different cultural picture with very few opportunities to articulate values related to the polis, jurisprudence, civil society, and eloquence. Stated differently, these values are rigidly codified on the basis of the principle of authority (Aristotle in physics, God in scripture, and the Church in theology). This is a golden opportunity for the triumph of dialectics, given that dogma must be defended with the instrument of reason against manipulation of that same reason by those who challenge the dogma. Instead, Humanism, as the word itself tells us, is above all a return of the human sciences and, therefore, of rhetoric. Here, however, we have a curious and significant lag, in the sense that the age of Humanism occurs in history at the same time as the arrival of a technological innovation, the Gutenberg printing press, which leads to the end of those communicative systems of the rhetorical type, i.e., based on the oral tradition, discourse, and verbal exchange conducted in person. This brings about the unprecedented triumph of analytical reason, but I have always warned that the consequences of technological change are never simultaneous. For all intents and purposes, the construction of a modern age was long and difficult, and the entire course of its development spanned more than 400 years. The analytical character of the Gutenberg model had an immediate homological equivalent (or a prelude) in the graphic, mathematical, artistic invention that was Renaissance perspective, but it did not have a similar homological equivalent at other "high" culture levels, marked by the reappearance of rhetoric (with the attendant complete faith in the centrality of discourse and the use of eloquence). In reality, the 1400s,

4 See my essay "La retorica di Cicerone" [The rhetoric of Cicero], in *Poetica e retorica* [Poetics and rhetoric] (Milan: Mursia, 1969).

1500s, and 1600s witness a series of alternating phases, characterized by faith in, followed by distrust of, the role played by rhetorical instruments; these are sometimes praised and at other times condemned in the name of analytical rigour. But undoubtedly, modernity sees increasingly the prevalence of the analytical, Cartesian, atomistic, and dissociating mindset. The geometric-mechanical model, with which the physical universe of the *res extensa* is interpreted, appears as the only reasonable and verifiable one. The human sciences either adapt to it or forfeit the privilege of being considered sciences, in which case they would be reduced to imprecise and confused disciplines. Among other reasons, this is due to the advent of electromagnetic technology (at the material level), to which Maxwell's equations for the electromagnetic field[5] correspond (at the "high" or ideal level); this completely changed the picture and heralded in the contemporary age. From that moment on, people reason at all levels using schemas based on continuity, the structurality of the field, and the dialectical relationship between the whole and the parts, rather than the sum of single elements. This means that, even in the physical-mathematical sciences, we find the fluidification of the foundational principles, with the attendant crisis of analytical reason and the re-emergence of dialectical-rhetorical reason. This allows the physical-mathematical sciences to be brought closer to the human sciences once again, thereby legitimating the attempt to establish a single methodology, for example, on the basis of such guiding principles as wholeness, structure, and *Gestalt* (which are broadly homological to the notion of electromagnetic "field").

5 James C. Maxwell, *A Treatise on Electricity and Magnetism* (1873; repr., Nabu Press, 2010). See in particular, the following remark, from page 81 of the introduction to the Italian translation, edited by Evandro Agazzi (Turin: UTET, 1973), which is perfectly homologous to what I state above: "Classical mechanics [had] fixed points and systems of points, a mechanics of the 'discrete,' which had simply 'adapted' its conceptual tools and its mathematical procedures to the treatment of the 'continuous' by considering limits, but the 'continuous' was always considered to be the sum (the integral) of its infinitesimal parts. [The laws of mechanics] permitted the assignment of the value of certain parameters in a given instant and relative to a certain point ... but did not express the general state of the system ... The equations of the electromagnetic field, on the contrary, are typical equations of a 'continuum' whose state is expressed by parameters that, in any instant, are taken to be finite for all the infinite points of the field."

It should also be mentioned, however, that there were flare-ups of analytical methodology in the various phases of neo-positivist or logical-analytical thought (which can be called neo-modern). For example, during the twentieth century, rhetoric experienced both low points and unexpected revivals. (These too were the effect of a kind of technological feedback: tape recordings or radio and television broadcasts of sound and image have enabled us to re-evaluate the behavioural, gestural, and acoustical elements upon which rhetoric and oral discourse relied during their heyday).

I do not intend to describe, even in general, the history of the logical-methodological or discourse disciplines. I simply wish to highlight their presence and influence, pointing out possible patterns of internal variation and, by means of the concept of homologies, link these variations to material culture. This would open up the possibility of re-classifying or ordering differently the various content disciplines that constitute the tree of knowledge of various historical epochs.

HOMOLOGIES BETWEEN THE ARTS AND THE SCIENCES

One of the main tasks of this entire study is to search for possible connections and interferences between one science and another. Such divisions and connections are governed by the formal or methodological disciplines, whose responsibility is to create the various units and to decide whether to bring these units into conflict with one another or to have them merge. In general, there are essentially three possible options. First, there can be peaceful coexistence or coordination between the block of physical-mathematical disciplines, which are grounded in analytical reason, and the social sciences, which are grounded in dialectical-rhetorical reason. In this case, each group of disciplines acknowledges the specific features of the other and does not consider it useful or economical to attempt to annex or expand imperialistically into the other by imposing its own methods and instruments on the adjacent block of disciplines. Such a wise coexistence is typical of complex and articulated cultural epochs that have developed in an equitable fashion. Evidently, the Athenian classical period was of this type since it witnessed the flowering of Aristotle's theories, which are indeed equitable and balanced. Our own era could also be considered of this type if for no other reason than its acute historical awareness, whereby it recognizes and appreciates the rights of each logical-discursive faculty and recognizes that it has never been possible to suppress or erase these faculties, whose unexpected resurgence has had to be acknowledged time and again. This could be the equivalent of a "soft" feature of our postmodern age, one that distinguishes it from the "rigid" features of more biased and unidirectional periods.

A second possibility presupposes the subordination of all the disciplines of knowledge in a given epoch to analytical reason and the logical-mathematical model. This has occurred at various times in the course of history and coincides with periods of expansion of mechanical technologies. The modern age, for example, has seen two vast periods of such imperialism on the part of the physical-mathematical sciences. One of these is in the second half of the seventeeth century, generally dominated by Cartesian thought and Rationalism (from the standpoint of McLuhan's theory). We should say that this is the moment in history when the Gutenberg technology ends its latency period or the phase of patient, subliminal shaping of the collective sensibility, an obvious equivalent being the philosophy of science at the "high" level of ideal culture. We also know, however, that the apparently contrasting intellectual climate of Anglo-Saxon Empiricism did not produce different results. In fact, our method of proceeding to search for homological correspondences is authorized to put aside the many superficial or cosmetic differences among the various schools, tendencies, and styles of thinking, in order to concentrate above all on the task of identifying deep modes of operation, which can often coincide or be homologous to a substantial degree in the same period. This does not mean that differences in expression are insignificant. There are undoubtedly some levels of study where these differences are important, but a general analysis conducted with the criteria of culturology can often do away with such superficial variants, just as in calculations it is possible to ignore the fourth and third decimal points in order to focus on the first and second. It is all a question of the level of analysis that one intends to perform.

Two centuries later, in the mid-nineteenth century, there is another "positive" phase where the physical-mathematical sciences triumph, the difference being that they no longer rely on a static model, but rather on a dynamic and evolving one. This is the product of a clear attempt to rebel against the "modern" domination of analytical reason and the natural sciences, attempted in various stages and with different rates of progression owing to the re-assessment of the historical-evolutionary nature of the forms or faculties of the humanistic disciplines. Dialectics and rhetoric imply the flow of time, unfold in a series of oppositional phases, and require intervention on the part of different subjects. They are fundamentally heterological and not tautological techniques, as is analytical thought, which is characterized by the ability to eliminate the temporal dimension and to unfold in a kind of achrony and simultaneity. Today analytical calculus is done by high-speed computers (whereas previously it was necessary to take into account the time required by scientists to perform the operations with the help of mechanical calculators). The fact remains, however, that the time to complete logical-analytical calculations, even when there is a measurable quantity, appears to be an extrinsic accident, whereas in the methodologies of dialectics and rhetoric it becomes an integral or organic part, an irreplaceable coefficient inseparable from their evolutionary nature.

Recalling what all the histories of philosophy tell us, however, the dialectical-evolutionary constructions of Romantic and Idealist thought at the end of the 1700s and the first decades of the 1800s are "naturalized" and undergo the well-known process of reversal, in materialist terms, perhaps because they are premature with respect to the state of the technology of the day. The concept of evolution is itself deprived of the human protagonist; stated differently, humans are seen as things alongside other things rather than as the subjects of a rhetorical-dialectical debate. Here, we have the triumph of the positivist phase of evolutionary naturalism in the modern age.

When we have similar phases of expansion, where one form of reasoning is imposed on the others in an attempt to subordinate them, what results is a process of reductionism. In other words, there is an attempt to reduce to the exigencies of analytical thought the field of human concerns (ethics, psychology, sociology, linguistics, etc.), to which the instruments of the rhetorical-dialectical group historically prove to be more suited. Chemistry, physics, biology, and physiology become the master sciences, and there is an attempt to reduce to that sphere disciplines considered to be composite or derived – for instance, psychology. To illustrate, the novelist Émile Zola attempts to present himself as a physician, an applied physiologist, and declares his program to extend the experimental method into the fluid and complex territory of human behaviour.[6] The humanistic disciples are faced with an alternative: either adapt to the process of reduction or be removed from the list of disciplines that can claim to conduct their work with scientific rigour. They then risk falling into the realm of evaluative or descriptive discourses and being considered imprecise and not very significant.[7]

Finally, there is a third option, which is the subordination of the various methodologies to rhetorical *ratio* or the logic of probability, fluidity, and the consensus of the community. Here too, in our search for general typologies, it is appropriate to recall specific historical periods when humanistic logic came to dominate. As noted above, in the Roman era and again in the time of Humanism and the Renaissance, this occurred as a result of the effective weakening of the disciplines of the physical-mathematical group, as well as the small number of changes in material technology, while social factors accentuated the importance of communal and political life. We could find something similar in contemporary (postmodern) culture but for profoundly different reasons. No one would claim that our age is characterized by a low level of technological innovation or by negligible development in physical-mathematical knowledge, when all indications are that there has been an unprecedented development in these two areas over the course of millennial human history. McLuhan's central theory, however, suggests that the basic technology provided by electrology (the application of electromagnetism and electrotechnics) corresponds to an epistemology or general methodology based on the prevalence of whole structures and universal features, rather than on quantity or summation. (This is the most important result of Maxwell's equations for the field of electromagnetism and subsequently of Einstein's theory of relativity[8]). On the other hand, the theories of probability and uncertainty formulated by Max

6 See the theorizing of Émile Zola in *The Experimental Novel* (New York: Cassell Publishing Co. c. 1893).

7 This bias re-emerged in the 1960s, in Charles Percy Snow's perhaps too successful work *The Two Cultures* (1959; repr., Cambridge: Cambridge University Press, 2012), where he downgrades humanistic culture. In this regard, see Preti's rebuttal in *Retorica e logica* [Rhetoric and logic] (Turin: Einaudi, 1968).

8 For an introduction to the theory of relativity, see Albert Einstein and Leopold Infeld's *The Evolution of Physics* (1938; repr., New York: Touchstone, 1967).

Planck, Niels Bohr, and Werner Heisenberg[9] prevail. The combined effect of these epistemological turning points is to bring the physical-mathematical group closer than ever before to the humanistic group, to the point where we can hope for a methodological unification that, for once, would not be the product of the dominance of one group by the other.

In many cultures these two distinct blocks emerge and gravitate toward either the physical-mathematical or the humanistic nucleus with their respective formal disciplines. At this point, it is appropriate to focus our attention on the arts, since this is my principal interest in the present study. We have already seen that the arts build a bridge with the "low" stratum of material culture. Unlike the sciences, they extend into the area of technology, even from the etymological standpoint, as the root word *art* indicates. The arts do not convey a body of knowledge or a system of notions and do not for the most part perform a didactic function (*docere*), but rather imply a doing or producing. A similar meaning applies to the "major" or nobler arts, also referred to as the fine arts, with which the arts have been identified through a filter of sorts provided initially during the Renaissance and then Romanticism. In the course of the modern era, painting, sculpture, literature, theatre, and music have increasingly accentuated their "inutility," which allows them to participate in the symbolic activities. This is not to say that they abandon their "performative" or technical aspect, which distinguishes them from the sciences. In a certain sense, they allow that aspect to atrophy or diminish, but they can never eliminate it entirely. Unlike the sciences, they do not communicate or transmit notions, but rather constitute objects, things that, even though loaded with symbolic values, go well beyond their mere physical properties. Furthermore, today no one wants to exacerbate the debate over the usefulness of the arts. Uselessness, that is, the spirituality of artistic activity, was a preoccupation during the Renaissance, when it was necessary to rescue artistic activity from the realm of manual labour, or during the proto-industrial era, when it was necessary to affirm that spirituality against stereotypical, mechanical pro-

9 A good account of the development of contemporary physics is *La neutralità impossibile* [The Impossibility of neutrality] (Milan: Bompiani, 1977) by Massimo A. Bonfantini and Marco Maccio, whose basic theories, however, I do not support.

duction. Nowadays, owing to a reassessment of technology (which above all demonstrates the fact that it can be articulated in various qualitative phases), the major arts have reconciled with their "minor" or applied sisters. As a result, the shame of "utility" has been overcome. If anything, its opposite sometimes comes to the fore, which is to say the shame of uselessness, the expectation of keeping one's hands clean. In this climate, the various intermediate fields (architecture, industrial design, popular literature, and the performing arts) are being strengthened. This is also true because uselessness is not measured simply in terms of productive utility or the production of notions, but also in terms of its implications for the vast field of entertainment and leisure, as noted in various periods by theorists who prescribe the function of *delectare* [to delight] as opposed to *docere* [to teach][10] as the objective of art.

Another feature that distinguishes the arts from the sciences (according to my own interests and corresponding to the meaning associated with the term) is a certain level of broad applicability; that is to say the sciences appear to be sectorial and have to carve out a slice of knowledge, or at least adopt a more selective filter. In this respect, the differences between sciences that deal with the external world or nature, and sciences that deal with man as a political, social, and cultural animal are important. As has been repeated several times, the first group generally falls within the domain of analytical reason, the others in the domain of rhetoric and dialectics. Although they have different methods, neither can function without analyzing and breaking up the continuum of reality, and without selecting from that continuum the specific objects of their respective studies. The arts, instead, tend to join, connect, and respect the highly organic and totalizing character of all human interventions. For this reason, they are similar to the formal sciences and the great logical-methodological systems that claim to have universal and transversal validity with respect to the various sectors. Hence, we have confirmation of the pertinence of the terminology. In the ancient and medieval world, these logical faculties (which permit us to speak of everything and, therefore, can be called discursive) were also called "arts"; thereby reconciling the two meanings on which I have reflected. They are not arts because

10 For these aspects of the historical debate on such categories, I refer the reader to various chapters in *Poetica e retorica* [Poetics and rhetoric] (Milan: Mursia, 1969).

they do not correspond to an accumulation of notions, but because they represent techniques or modalities for working on things, and they have broad applicability.

The disciplines that have concerned themselves with the arts (poetics, rhetoric, and aesthetics) in the various epochs have more or less underlined these features, namely, the fact that notions are incorporated into things in the artistic sphere and they cannot be separated from those things with impunity. The two are linked organically together in a given context. Through the writings of Susanne Langer, the school of Ernst Cassirer, which more than any other emphasized the notion of symbol, conveniently introduced a distinction between a discursive symbolism that is peculiar to the sciences and a representational symbolism that is appropriate to the arts.[11]

CULTURAL "OPERATORS" AND THEIR LONG-DISTANCE COLLABORATION

It was appropriate for me to recall all these things in order to avoid creating the impression that I wanted to oversimplify or to turn the arts and sciences into a sort of featureless no-man's land covered by waves of uniformity, as when a river overflows and floods the living and complex geography of the surrounding territory with a monotonous, muddy surface. Such an image corresponds to epochs in which the error of reductionism was committed, as suggested above; this is the attempt to impose the logic of one group of disciplines, or indeed of a single discipline, on even the most remote sister disciplines. It should be clear that reductionism has been an important phenomenon that occurred in certain historical periods with such legitimacy and utility as to produce substantial progress in human knowledge as well as a better organization of that knowledge. This was the case in the second half of the 1600s and the mid-1800s. Today, reductionism undoubtedly gets some bad press, even if a new attempt at annexation launched by analytic reason, assisted by linguistics and semiotics has only recently been repelled. As noted above, a push in the opposite

11 See Langer's *Philosophy in a New Key* (1942; repr., Cambridge, MA: Harvard University Press, 1979). For a comment on Langer's philosophy, I refer the reader to the study in my book *Per un'estetica mondana* [Toward a worldly aesthetics] (Bologna: Il Mulino, 1964).

direction has occurred, originating in the "soft" and permissive logic of the rhetorical group of disciplines, a fact that tends to confirm the liberalist tendency of our age.

But if reductionism, which is to say the prompt and easy confluence of every specific sectorial feature into one confused solution, can be a negative, at least in many periods of history, so too can its opposite, sectorialism, or the excessively protectionist stance of disciplinary autonomy, and the excessive pseudo-scientific reasoning used to defend it. In this second case, supposedly watertight bulwarks would appear, isolating the individual fields irreparably. The analogy of the communicating vessels is once again suited to our case in as much as it does not presuppose the uniformity of the vessels; each can maintain its own particular form and capacity. (Not speaking metaphorically, this would promote respect for the sectorial specificities and the autonomy of disciplines). However, this does not prevent the circulating liquid from attaining the same level in all the vessels.

Naturally, this is only an analogy that undoubtedly contains some common-sense value, but it is largely imperfect. In the physical world, the attainment of the same level is almost instantaneous and occurs naturally; the observer needs only to register the event with his or her eyes. In culture, however, the verification of the fact that the same level has been reached in different sectors is itself the product of technical work. In other words, it is not a self-evident fact, but one that must be generated through bold and risky operations. This is also true because the difference in vessels (to return to the analogy) is more significant here. The variety of shapes of the sectors and the different ways in which the "specific" units interact can be a major obstacle to the identification of homologies (the fact that the same logic is working in the group of sectors being examined). It can also happen that defence mechanisms appear; these are mechanisms disguised as plausible reasoning, which relies on philological rigour and scientific integrity and seems to argue that it is impossible to make comparisons between sciences because they are independent of one another. What ensues is a dangerous bias according to which such comparisons between sectors are legitimate only if there is historical proof of reciprocal influence; that is, if it can be shown that the "operators" in one sector were really aware of the accomplishments attained in another sector. It would be legitimate to correlate a representational or symbolic system like painting and a scientific field like physics only if the artist in question had "really" studied the theories produced in that other area.[12] But here we would lapse into reductionism because we would be giving priority to one particular area of inquiry; we would be choosing a kind of monogenetic hypothesis as though there existed one vessel, a special vessel through which the new liquid (metaphorically speaking) would circulate. The circulation would occur in one direction only, from the primary vessel to the others, which are connected either diachronically or synchronically. In this scenario, the fact that we find the same functional paradigms in adjacent areas would lose a good part of its explanatory value and become a tautological consequence.

If, instead, there are no philological documents to prove reciprocal influences among the "operators" in two different fields, the identification of homologies in the way of thinking and working acquires great explanatory power. The respective achievements corroborate one another, providing what could be called a cross-verification. It is not a question of derivation but of simultaneous accomplishments. This does not mean that there are no connections, but the clear evidence of a common level attained (of homological correspondence) becomes the main reason why we can search for such points of contact. This is what happens with the communicating vessels. If rainwater reaches the same level in two caves, it means that they are connected. All we need to do is to find the passageway that joins these caves. To speak of the same level reached in different symbolic cultural environments, however, is to use a simile or an analogy. We need to work very hard to confirm the existence of correspondences and to have the courage to go beyond the specific contours of an individual vessel (the powerful and valid *raison d'être* of each discipline). Furthermore, there are usually lags. The various arts and sciences reach at different rates, some earlier and some later, the point of synchrony, the attainment of the same level in terms of method and function. In certain historical periods, some arts and sciences play a leading role

12 I have examined and refuted these types of biases on at least two occasions: comparing "operators" in the field of literature, such as Svevo and Pirandello, with a philosopher, Bergson, and a scientist, Freud (exponents of rhetorical, dialectical, and analytical logic, respectively) in *La linea Svevo-Pirandello* [The Svevo-Pirandello line] (Milan: Mondadori, 2003); then comparing a philosopher, Bergson, and a linguist, Ferdinand de Saussure, in *Alcuni problemi epistemologici relativi al 'Corso di linguistica generale'* [Epistemological problems in *Course in General Linguistics*], in *Lingua e stile*, 3 (1968).

while others allow themselves to be pulled along by the momentum. As well, it cannot be said that the circulation occurs in all the vessels. The various blocks or disciplines could remain distinct and not communicate with other disciplines or vessels.

Thus far I have been speaking abstractly about sciences, disciplines, and arts, leaving aside the fact that they are made functional by their respective "operators." I would like to point out the appropriateness of using such a deliberately generic and decontextualized term as "cultural operator" because it has the advantage of establishing a lowest common denominator among many different professions, like those of artist, critic, scientist, philosopher, etc. This does not mean that their respective responsibilities are erased. Each has to continue to respect the norms of his or her own sector, but this is not to deny the opportunity to attain a common level, which would provide the best demonstration of the fact that the vessels communicate with one another despite the clear individuality of each and they participate in the general circulation of ideas. Furthermore, the "operators" within the different disciplines are certainly not equivalent or interchangeable. Some defend and ensure the *status quo*, while others become agents of innovation and experimentation. A culturological approach to the subject, however, certainly does not praise individualism. It is very difficult to imagine that an invention, a discovery, an artistic, scientific, technological, or methodological innovation is the exclusive product of the genius of a single, extremely talented individual. Conversely, it is difficult to imagine that a discovery is an isolated event, without echoes, correspondences, or homologies. On the contrary, we can demonstrate that each has an equivalent in other sectors that may be located far apart from the standpoint of logic (within the encyclopedia of a period) or geography (that is to say, emerging from countries with different ethnic and linguistic traditions). The only distance not admissible is a chronological one; at least, if one is permitted, it can allow us to acknowledge the great lag between one sector and others that are more productive and active.

This means that the conditions favouring the appearance of an innovation are objective. They result from the common practice of human community, as it occurs in a given epoch, giving birth to a precise historical culture: both material and ideal, both real and symbolic praxis. So-called individual geniuses do not create the conditions in or for themselves, but grasp them with greater speed than others do and accelerate their maturation. They make them "precipitate" more rapidly, but in so doing merely allow us to save a bit of time since others would have reached the same conclusions in other geographic areas or other disciplines. In sum, the individual performs a function that one of the most authoritative founders of culturology, Leslie White, has compared to that of catalysts in chemistry. These substances do not enter into the atomic structure of the synthesis, but they accelerate or retard, enable or prevent, these syntheses.[13] Such a reduction of the role of the individual leads me to favour the hypothesis of a polygenesis of innovations. It is difficult to imagine that these arise from a single hot bed. In general, we can find several sources with respect to diversity and distance, both geographic and disciplinary. Only in this way can different scientific results reinforce one another and submit to a useful process of cross-verification.

13 See Leslie White's *The Science of Culture: A Study of Man and Civilization* (1949; repr., New York: Farrar, Strauss and Giroux, 1969), 146–89.

3

The Internal Articulations of Cultural Systems

THE SEARCH FOR THE NEW

In the preceding chapters, I reflected on the relationships that the visual, literary, and performing arts have with other aspects of a given culture as well as on the factors that cause the arts to change in tandem with those aspects of culture. More to the point, I have discussed three types of relationships: vertical relationships, in which artistic activity and critical or historiographic study of that activity establish with the so-called material stratum of culture or technology (the implements of production, transportation, and work in general); and horizontal relationships, which link the sphere of the arts with other spheres of "high" culture, characterized by the use of ideas as intellectual instruments (what we tend to think of as culture per se, which is really an oversimplification). This stratum encompasses the work of various sciences, especially in their theory or conceptualization stage since they participate in material culture when they enter the application phase. What is left for me to do is examine a third type of relationship, one that concerns the internal relationships, criteria, or factors that explain the variations art produces within its own realm and that do not have immediate or direct equivalents in other areas of inquiry in "high" culture or "low" material-technological culture. It is worthwhile specifying, however, that culture is an organic and synergetic whole and, therefore, the distinction made among these three types of relationships is a useful abstraction that allows us to examine the components of the mechanism in a stationary state, so to speak, keeping in mind that they function in a strictly interconnected way within the dynamic of culture.

The first factor of internal change in the arts can be identified as the unending search for the new. Innovation certainly does not appear only in the arts and in aesthetics; it is a basic feature of human culture.[1] The "cultivation" of fields and the environment in general by human beings ought to be distinguished from that performed by animals precisely because humans are able to renew their implements and increase their efficiency, thereby satisfying a primary biological impulse to expand the use of one's faculties and increase the level of one's well-being. Consequently, the search for the new is at the root of the human condition and is part of what makes that condition unique, giving humanity an advantage over animals. Therefore, we cannot attribute purely formal characteristics to that condition.

If it is true that this tendency to renew implements is common to all cultural activities, whether "low" or "high," it is also true that, when it comes to artistic or aesthetic activities, we find an unparalleled emphasis on it as well as the trend to transform itself from a means to an end. In the arts, the search for the new for the purpose of improving practical and theoretical solutions to problems and avoiding impasses becomes an end in itself. It may be that this operational sphere is directed toward cultivating the precious human faculty of innovation and toward developing a kind of perpetual training in this field. In short, in the practice of art, innovation no longer appears as content to be acquired, but rather as form, a moral imperative, a model for all our actions and experiences.

In fact, it is possible to say that the category of the new has appeared frequently in definitions of art and aesthetics in various cultural epochs. There is no doubt that this is a recurring feature, even if we should guard against drawing from this fact an absolute definition for this characteristic of innovation. This would be to assume that it must always and in equal measure become part of a cluster of fixed traits. On the contrary, we can immediately find an element of changeability in this feature. In the course of history, there have been epochs characterized by this tendency to innovate and others that were cold and cautious, or inhibited in this regard. We may be able to concede that

1 From this point forward, I use artistic and aesthetic as a hyphenated term to indicate practices involving the production techniques for a material object and interventions designed to enliven or stimulate sensibility, respectively. At the same time, joining the terms indicates a desire to create a single block of the practical and theoretical concerns (aesthetics and poetics).

our own era clearly belongs in the first category; this permits us to extrapolate the new as a formal principle that is immanent to the field. However, the search for the new in a pure state that pertains to art does not exclude the possibility that a similar search for innovation can be found in the other sectors of culture.

THE GENERATIONAL CONCEPT

This form or immanent principle, which is so much a property of the artistic-aesthetic field, has in the concept of generation its typical "vehicle" that does not have an equally obvious equivalent in the other sectors of the culture. This criterion is, first and foremost, biophysiological; that is to say it applies to those born in a relatively narrow window of time and is distinguishable from a preceding similar period by the fact that it occurs about twenty-five years later, which is the timespan of a generation or the temporal distance that separates parents from their children. If it is initially biological in nature, however, the generation loses that feature the moment it appears at the cultural level. In fact, biological differences between generations are not particularly interesting in themselves. (After all, even common sense tells us that these differences are negligible since biological differences can be measured only across long periods of time.) Much more important are psychological differences as well as psychoanalytic differences, following Freud's theories, which suggest there exists not only a relationship of affection and peaceful transmission of values from parents to children, but one of conflict. Fathers want to "eliminate" their sons and prevent them from enjoying the sources of sexual pleasure as well as other forms of power and pleasure. Sons want to kill their fathers, even if only symbolically, in order to take their place. To this point, we have the Freudian "myth," a meta-historical one that nonetheless has important explanatory power with respect to historical struggles between generations over control of the centres of power, administration of public affairs, and choice of lifestyles. In all this, the category of the new plays a significant role since it is generally possible to delineate the two sides of the public drama where the older generation is characterized by a certain misoneism (hatred of the new) along with the preservation of political, intellectual, and aesthetic power, whereas the younger generation is characterized by philoneism (excessive love of the new).

But, if this generational concept and the extremes of attachment to and aversion for the new underlie the entire domain of culture, it is precisely in the artistic-aesthetic sector that these polarities manifest themselves with greatest clarity. In all other areas, in fact, there is the need to come to terms with the immediate, real, and material factors of renewal. This is especially true in technology, but it is also true in the physical-mathematical sciences and in the social sciences. It is not enough to "want" to replace the preceding generation and confidently play the innovation card. This innovation has to produce a certain amount of practical reality and not merely casts of mind. However, the importance of the desire to innovate as an end in itself is not to be undervalued even within the sphere of material culture.

When we come to artistic production and the activities related to it, the impetus toward the new is more justifiable for the reasons indicated above, and this makes it a determining condition that explains generational revolutions. It becomes almost obligatory for each artistic generation to perform the ritual sacrifice of the father; otherwise, there would be no room for the generation to establish itself and to assert its own reason for existing. Naturally, here too, there is a reciprocal situation; that is to say, as in the logic of other fields of inquiry, there are not only material, external factors at play, but also formal, internal ones (the push toward the new as an end to itself). On the other side, these formal impulses have to come to terms with external factors. In other words, generations do not always succeed in establishing themselves. Indeed, some generations are sacrificed, not so much for lack of intrinsic talent (after all, this is a gift that is difficult to evaluate in itself) as for lack of favourable external circumstances. It is also because the older generation has brought to the table many innovations, which the younger is unable to equal. We know that, when an artistic generation emerges in synchrony with the great technological, epistemological, and scientific changes in a culture as a whole, it is easier for that generation to achieve solid results, but it would be naive and unprovable to claim that generational changes at the formal level always have an equivalent in the infrastructure of human society.

Even a modestly careful historiographic analysis tends to identify what I prefer to call a "nebula of statistical density," which means that the protagonists of any "ism," movement, or tendency appear to be born within a certain window of time, provided we conduct the analysis in statistical terms and do not look for absolute answers. The Impressionists are born "around 1840," the Symbolists in painting and literature "around the 1860s," the Cubists and Futurists "around 1880," the Informal artists "in the first decade of 1900s," and the generation that invents Pop Art is born between 1930 and 1935. This is also true for those who developed the literary Neo-avant-garde in Italy and France. Such an analysis should be extended to other periods and conducted with mathematical-statistical techniques to determine if these observations can be confirmed or if, instead, these groupings are artificial products of critics who have a clear idea of what they want to prove and, therefore, skew the data to suit their purposes. It would be surprising if a rigorous analysis did not confirm the logic of the generational paradigm since it appears to be so powerful and plausible, supported as it is by many factors that are not biological-physiological, but rather cultural. These include being exposed to the same influences, the same readings, being impacted by the same external events, the possibility of contact and direct collaboration, which often prompts the production of manifestos and statements of poetics, the creation of schools and groups, and the founding of militant journals. These are all forms of aggregation marked by the compactness and homogeneity of the generation, even if we cannot rule out the possibility of an intellectual or even personal encounter with a member of a preceding generation, someone in a position to break with his or her contemporaries.

But the generational paradigm is even more powerful than these concrete manifestations of groups or associations and, in this sense, it presents particular responsibilities for the historian and the critic. These people should not bow to the pressures exerted by philologists obsessed with direct relationships; in other words, generational homogeneity does not occur only in the case of those who acknowledge the existence of that homogeneity by effectively forming groups that produce manifestos or other types of organizations. It may be that the producers of manifestos do not know one another or may be adversaries; a careful examination of the principles underlying the various declarations may reveal a substantial affinity or homogeneity. Explicit declarations, positions consciously taken, and affiliations are certainly a valuable point of reference for those wishing to understand an epoch, but it would be a mistake to consider these to be definitive and necessary elements. The circulation of cultural ideas may have been more vast and more transversal than appeared at first sight, and the need to react to the choices made by the preceding generation may have been a more vital and unifying factor than believed. In this scenario, the members of the same generation, although dispersed geographically, may have attained a common level of sensibility or have made the same stylistic choices. Here again an analogy can be drawn with the communicating vessels phenomenon, which occurs even when the channels of communication are sometimes not very evident or appear to be of very different dimensions; although some are larger than others and placed at substantial distances from one another, the liquid flows and reaches the same level throughout.

Naturally, the fact that a common level has been attained is the best proof that passageways must exist. This can be said to be analogous to the stylistic unity and compactness attained by the various representatives of the same generation. The fact that we can easily identify the feature serves as the stimulus to look for passageways and secret connecting channels. This contrasts with the philological mindset, which allows us to maintain that relationships exist only where there are explicit and demonstrable signs of connection or derivation. Finally we are led to adopt the notion of the polygenesis,[2] rather than the monogenesis, of styles. There is hardly ever a single diffusion source; instead, there are many distinct sources that may seem not to be communicating with one another, and yet there is communication. The proof is the attainment of the same level, which is, therefore, a circumstance that becomes a heuristic instrument in the search for supporting philological evidence (since we must reject the possibility that stylistic uniformity is achieved by magic or spores carried in the air). It is also worthwhile repeating that a basic argument in favour of polygenesis in the absence of "tangible" contact points is the fact that the need to go beyond the achievements of the preceding generation is felt everywhere. Even in the absence of other connections, this common negative criterion would be sufficient to allow us to understand the unavoidability of "marching together." As we will see more fully in the paragraphs that follow, this is because so few possibilities present themselves to those who are convinced of the need to reject a certain style and allow the emergence of a new style in order to establish a

2 The spread, at least in western culture, of phenomena like the Baroque (Conceptism, Gongorism, Préciosité, "ingegno" or wit, "Metaphysical Poetry," etc.) or more recently Symbolism (Art Nouveau, *Jugendstil*, Liberty, Modernism, Decadentism, etc.), as well as the Informal (Tachisme, Art Autre, Abstraction Lyrique, Action Painting, Concrete Art, Gestural Art, etc.) can only be explained in terms of polygenesis.

generational identity; therefore, cultural "operators" who may be distant from one another and not communicating with one another are obliged to find solutions that are close at hand or perhaps coinciding.

The generational paradigm is also confirmed *e contrario*, which is to say through the "exceptions that confirm the rule" or pieces that do not fit. In fact, we can say that life is not easy for those whose date of birth falls "outside the generation," outside what I have defined as the "nebula of statistical density," which is to be understood as having a certain amount of elasticity because a lag of a few years is not a significant deviation, unlike a difference of ten or more years. Here are some examples: Impressionist painters Alfred Sisley and Pierre-Auguste Renoir, born respectively in 1839 and 1842, fall perfectly within the window (if we take the median to be 1840). The same can be said for Symbolist painters Giovanni Segantini (1858), Georges Seurat (1859), James Ensor (1860), and Edvard Munch (1863); their median date of birth is 1860, exactly twenty years after the corresponding date for Impressionists. But this is not the case for Paul Gauguin, who is born in 1848; hence, too late to be a full-fledged Impressionist and too early for him to acquire a clear and self-conscious anti-Impressionist stance. In this case, the distance from his older contemporaries rather than the previous generation is long and tormented, as is the attainment of harmony with his younger contemporaries, Émile Bernard and Louis Anquetin. Another dramatic date of birth falling outside the generational window is that of Henri Matisse, that date being 1869; therefore, he is born too late to be a Symbolist and too early to detach himself completely from that climate. Until 1905, he is influenced by the Divisionism of Seurat and the Synthetism of Gauguin before finding his way; after all, he was profoundly conditioned by fin-de-siècle two-dimensional painting, never accepted the new canons of Cubism, and was overtaken by the young Pablo Picasso and Georges Braque in the race to participate in the more audacious exploits of the avant-gardes. Here too, it would be useful to conduct a careful statistical analysis to determine if these first glaring examples are confirmed by a much broader sampling.

PATTERNS OF QUANTITATIVE VARIATION

If we accept the assumption that the main responsibility for the quest for the new, or at least the cause of changes in style, lies with the generation, what remains is to establish what types of relationships exist among the various generations: continuity, opposition, rupture, advancement, intensification, etc. To avoid ambiguity, I would like to say from the start that such "types" or patterns of variation are the typologies I have adopted for descriptive convenience and with no expectation that they would be "substantial." They do not, however, lack an amount of interpretative value. Essentially, every generation is different from every other. History itself is the well-known river whose waters never wash the same place twice. Furthermore, the most substantial variations are the ones that emerge in tandem with turning points that occur in other sectors of culture, beginning with the "low" or infrastructural sector of technology. In the interests of clarity and economy, this does not preclude the desirability, or even the necessity, of using a limited number of schemas, provided they are tailored to each phenomenon (each generation) under examination, especially since our imagination (perhaps even reality itself) is capable of producing only a limited number of patterns. Formal patterns of variation can be sub-divided into two categories, depending on whether they involve the quantity or the quality of the variations. In the first category are those patterns that tell us nothing about the choices themselves, for which a new generation assumes responsibility; but these same choices characterize the quantity of the differences in their work relative to that of the preceding generation. In the second category, instead, the question of the quality of the variations is an issue, although it is always at the level of general morphology or broad frameworks that need to be filled with custom-made content.

For the quantitative patterns of variation, we can fish out the old concept of the "transition period," which appears so frequently in art history books. It describes very well generations that find themselves suspended between the old and the new, having already rejected some things, but not having the capacity to adopt perspectives that enable them to go beyond a certain barrier. Proof of the appropriateness of proposing such an intermediate stage is the fact that it would not otherwise be possible to locate new experimentation entirely within the old period or to place it completely within the new; an example would be Symbolism in both painting (Georges Seurat, Paul Gauguin, the Nabis, Gaetano Previati, and Giovanni Segantini) and in literature (Maurice Maeterlinck, Gustave Kahn, Oscar Wilde, Giovanni Pascoli, and Gabriele D'Annunzio). On the one hand, these cultural "operators" no longer have faith in nature and science, which is typical of the second half of the 1800s. They sense the radical insufficiency of both; on the other hand, they find it difficult to adopt the attitude that would typify the first decades of the twentieth century, i.e., full awareness of the fact that the drama unfolds within us, that it is we, with our stratified consciousness, who exert will or do not, we who "construct" nature and science, shaping them to conform to our own epistemologies. The compromise of the Symbolists is the fact that they objected to nineteenth-century positivist thought (nature as the source of all certainty). At the same time, they retained the tendency to look outside the self for answers, thereby replacing nature with another external entity, a supra- or sub-nature that both produces and justifies mystery, enigma, and the uncanny in the same incontrovertible way that positivist nature produced binding "laws." This is neither the 1800s nor the 1900s but a transitional phase with its ingenuous and ingenious *bricolages* that put old and new materials in a special equilibrium, which can be understood only if we take into consideration the need for compromise; otherwise, if approached with the univocal interpretive standards of the preceding epoch, such an equilibrium would be condemned as indeed it was by nineteenth-century Naturalists and *Veristi* (Italian Realists) as well by critics from the "historical school" who disapproved of the spiritual degeneration of the "Decadents," or by Futurists, Expressionists, and Dadaists, who did not want to recognize them at all as their immediate predecessors.

The notion of "transition," therefore, seems to emerge from less than the full ripeness of time, which is to say, when the results produced by the dominant generations in a particular historical moment are no longer satisfactory, but there is not a sufficient accumulation of factors, impulses, and contributions from other fields of inquiry to enable radically different styles to "precipitate." If these indicators of maturity appear in great numbers, we can have almost the opposite phenomenon: a violent and sudden rupture, the emergence of a rebellious generation fuelled by a sacred combative fire, by an iconoclastic spirit, and by all the other traits of avant-gardism, even those external to it. We can draw an interesting analogy with what occurs in geophysical phenomena where, as abstract logic would suggest, the highest mountain chains are not located at the centre of continents, at the point of greatest distance from the oceanic trenches, where they can rise to greater heights slowly and gradually. On the contrary, the highest peaks rise mostly near valleys in dramatic juxtaposition and are perhaps related dialectically to the same process. This is also the case when the time is ripe for stylistic ruptures. These can occur with sudden violence, which is not subsequently equalled without undergoing a process of gradual development, one step at a time, toward the apex of bold exploits, which should be found at an appropriate temporal distance from the fault line. By contrast, avant-garde movements appear rapidly, reach their apex, and then begin to decline. The situation following the explosion is generally milder and gentler than that immediately preceding the explosion. The "roaring" avant-gardes of the 1910s and the 1920s, especially Futurism and Dadaism, expressed impetuosity and proposed alternative lifestyles charged with a rebelliousness that no phenomenon can match, much less surpass, even today. One of the consequences of this revolutionary harshness is the development of a powerful conflict between the few initiates or *illuminati* and the others, the Philistines, those who remain in squalid normality.

The destiny of generations that come after these trailblazers, the first to ascend the peak, to scale the Himalayas of revolution, is more difficult and unrewarding. In many cases, there is nothing left for them to do but take the path that leads to "normalization." Resorting once more to a useful geophysical analogy, this is a matter of "levelling" emerging phenomena, but with the advantage of popularizing them and transforming the mountain chains and peaks into stable, flat plateaus that extend over vast areas. In fact, "normalization" contains an element of quantification and even democratic "proselytism." It is necessary to spread the heroic innovations of the preceding generations in order to make them accepted by the majority and to eliminate their power to shock and scandalize. I found it indispensable to apply this schema to my own studies on the so-called "Italian Neo-avant-garde" (the generation born around 1930, the *Novissimi* poets and their critics),[3] who found themselves compelled to re-propose features that had already appeared in the works of Dadaists, Surrealists, analytical narrators like Joyce, as well as works by Italian Expressionists and Futurists of the period of the Florentine journal *La Voce*. These movements, however, were justified by the fact that they were quantitative re-presentations, instances of the expansion of the system and homogenous diffusion. They occupied vast areas as opposed to being flashes of revolutionary

3 See, for example, *La normalità "autre" di Sanguineti* [The "odd" normality of Sanguineti], reproduced in the anthology *Gruppo 63, critica e teoria. Dalla nascita del "Verri" alla fine del "Quindici"* [Gruppo 63, criticism and theory, from the birth of *Il Verri* to the end of *Quindici*] (Bologna: Il Mulino, 1995).

action, and too bad if this image of a "normalizing" generation gets bad press, if it pleases fewer people and seems less "beautiful" than the image of a heroic avant-garde disdainful of the "ordinary" person. We need to realize that cultural "operators" do not choose what they want to be in the abstract or in absolute terms; instead, they have to adapt to the concrete circumstances that present themselves, in keeping with the generational paradigm. As a result, they are strongly conditioned by what was done by those who came before (on the internal side, because there are also crucial external conditioning factors). On the other hand, even if it is not especially eye-catching, the task of "normalization" is also important. In fact, the transition from quality to quantity does not occur automatically or by inertia; rather, it entails bold solutions with respect to structure and precise technical inventions. We need only consider a typical "normalizing" phenomenon like the *nouveau roman* (Alain Robbe-Grillet, Michel Butor, Claude Simon), which established itself so successfully worldwide. No one can claim that these novels offered stylistic features radically different from those of James Joyce, Franz Kafka, and Virginia Woolf, but no one can accuse them of passive and stale repetition either.

It is also impossible to avoid "normalization" in the experiments that characterized the visual arts of the 1960s, when there was a systematic quantitative revival of all the typical Dadaist techniques: ready-mades, conceptual word-games, collages, photomontages, photograms, etc. In this case, quantity is tangible since the procedures, environments, behaviours, and performances of this aspect of "normalizing" Dadaism occupy much greater spatial and temporal areas than did its historical teachers, while at the same time the intent to scandalize is diminished. In fact, a didactic process of aesthetic education is triggered, aimed at persuading the public that "art is easy" and within reach of everyone.

"Normalizations" also have the task of inventing structures and are not phases of regression but consolidation and extension of previous achievements. They should not be confused with the "call to order" movements. Having considered the contributions of preceding heroic generations, the members of the first group (those who "normalize") realize that they have no room for qualitative variation, but see before them the vast amount of work they can accomplish within the program already laid out for them. The members of this second group (those in favour of a "call to order"), instead, understand that there is no room, either qualitative or quantitative, or are at least aware that insisting on innovation would only lead to modest variants, would not open up new avenues, and would oblige them simply to follow in the footsteps of their predecessors. So they are compelled to adopt the opposite strategy, that of searching in the storehouse of traditional ideas and forms disdainfully rejected by the avant-gardes, and discovering fascinating aspects in that same storehouse. In short, it is a question of playing the "novelty of tradition" card against the "tradition of novelty" card, which appears to characterize the "normalizing" cohort. All this occurs in concert with more or less direct socio-cultural changes that justify a period of retreat and reconsideration, but these justifications may also not exist. The mania for order could also emerge from purely internal necessity, as a dynamic impulse or a swing of the pendulum that restores equilibrium. The most spectacular self-named "call to order" movement occurred in the 1920s and it certainly enjoyed the external support of social involution. (Let us not forget, however, that it affected all the countries of the western world, many of which did not experience the phenomenon of Nazi or Fascist dictatorship). Swings of this type, however, have been repeated recently as well within an external framework that does not exhibit evident or univocal symptoms of recession and involution. The fact is that even today, having recently emerged from (or possibly being still immersed in) the practices of normalized Dadaism, which was devoted to the "heroic" avant-gardes (Futurism and Dadaism), we quickly discover a fascination with the 1920s and 1930s, Art Déco, Metaphysical Painting, *Valori Plastici*, figurative mural paintings, and revivalism in general.

PATTERNS OF QUALITATIVE VARIATION

By speaking of the "call to order" movement, we have shifted from patterns of variation that I have called quantitative to those that are qualitative in nature and as such say something about the styles of a given historical period, even if what it says is always general. Although vague, the concept of "order" is meaningful and evokes a certain set of values as opposed to others. In fact, it is always confused with the concept of "closed," as described by the most famous theoretician of such patterns of morphological variables, the art historian Heinrich Wölfflin, who is famous for proposing the binary opposition "open-closed."[4]

4 See Chapter 6.

As is well known, Wölfflin did not propose only this binary model, but suggested others as well (linear-pictorial, surface-depth, multiplicity-unity, and clarity-obscurity). These additional pairs, however, appear to be more pertinent than the first to a specific area of study, i.e., the visual arts, therefore they suffer from a more limited range of applications. "Open" and "closed," instead, are notions that lend themselves to very broad applications far removed from their denotative meaning. They are rich in metaphorical meaning, which allows us to apply them to all the arts, not just painting, but architecture, music, the performing arts, and literature as well. We need only think of how successfully and broadly Umberto Eco uses these terms to develop his notion of the "open work." There is no reason why we should limit the use of Wölfflin's terminology to the aesthetic field. Indeed, in common usage "closed" and "open" are applied to psychological traits. (We speak of a "closed" or an "open" personality, for instance.) These two terms, therefore, can contribute to realizing the goal of merging the sphere of artistic phenomena with other spheres that pertain either to "low" material culture or "high" conceptual culture.

These same excellent opportunities for application also concern the "high" culture disciplines of epistemology and psychology, including the subsequent development of psychoanalysis. The epistemology of the Galileo-Newton era can justifiably be called "closed" because of its faith in the human capacity to anticipate and to measure the properties of physical phenomenon. By contrast, we can define as "open" the era that accepts the Uncertainty Principle proposed by Werner Heisenberg and especially the relativism introduced in the existential philosophies and the notion of contingency, developed toward the end of the nineteenth century. Lastly, in terms of the psychoanalytic interpretation of history and great collective phenomena, we can identify a model of "closed" behaviours where the Ego, or perhaps the Super-ego, is dominant, and a model of "open" behaviours where the manifestations of the Id, the Libido, are more numerous and the forces of repression diminish.

The "closed-open" dichotomy, therefore, has broad applicability since it can be used at the purely internal level to designate nothing other than styles and the morphological configuration of their surface features. It can also have a synergetic relationship and become part of the corresponding features that appear as shifts in technology, turning points in epistemologies, or broad shifts in the Ego/Id equilibrium, i.e., the "pleasure principle" and the "reality principle." Such a bipartite schema in itself is merely an empty frame, an outline that has to be filled with content provided by specific contexts that are still strictly stylistic. These may still be strictly stylistic, identified through careful examination of the works themselves, and consequently placed in one category or the other, but they may also emerge from a comparison with the features of both the material culture and those belonging to the other sciences.

In the end, although it is the most well-known and perhaps the most easily verified, this binary system is certainly not the only one. Many others can be formulated from time to time in response to the need to find appropriate instruments for an area of inquiry, as long as we are fully aware that such variations are not inscribed in the nature of things, but are merely convenient human constructs designed to synthesize data and convert that data into effective models of understanding. I would like to conclude this section by recalling a very useful pair of terms in the history of narrative forms, originating in Aristotle's classic treatise, *Poetics*, where the author states that in the first two parts of every long poem (epic, tragedy, comedy, which is to say, the forerunners of narrative genres still practised today) *mythos* and *ethos* emerge: actions (plot, adventure) and character (the psychological or sociological density of the character).

Therefore, we find phases of narrative experimentation and production designed primarily to satisfy the exigencies of the first component (action or the pleasure of narrating above all, which implies, by contrast, thinness of characters, lack of psychological depth, and reduced value of the setting) and others that, in response to the first, focus on the characters in order to understand their motivations and tend to downplay the structure of the plot. This is another abstract schema of pure variability that certainly needs to be verified and filled with concrete facts.

4

The Birth and Evolution of "Modern" Space

PERSPECTIVE AND TYPOGRAPHY

On several occasions, I have observed that, at the onset of the modern era, we find a typical homology between an invention pertaining to material culture or technology and one that can be located at the intersection of various disciplinary fields of knowledge or "high" culture, typical of symbolic systems. One such invention is the movable type printing press or typography, generally attributed to Johannes Gutenberg and dated around 1450.[1] The other is so-called Renaissance perspective, which allows us to confirm the existence of a substantial homology between the sciences and the arts (painting, mathematics, and philosophy). The scholar who described this convergence with the greatest lucidity was Panofsky,[2] while the printing press phenomenon was studied especially by McLuhan. It should come as no surprise that the homology "internal" to the sphere of sciences was noted before the other. Panofsky's famous essay, in which the author affirms the symbolic form that pertains to perspective, is from 1927, while we have to wait almost a half-century before the Canadian scholar ventures to compare the two levels, which are apparently

1 For the history of typography, a very useful work is Lucien Febvre and Henri-Jean Martin's *The Coming of the Book* (1958; repr., New York: Verso, 2010). To this we can now add the monumental volume by Elizabeth L. Eisenstein, *The Printing Revolution in Early Modern Europe* (1979; repr., New York: Cambridge University Press, 2012).

2 In *Perspective as Symbolic Form*, (1927; repr., Cambridge, MA: MIT Press, 1991).

distinct and separate from each other; namely, technology and sciences. Perhaps today there are still some who tend to react negatively to such an attempt and deny its legitimacy.

The first theorization of perspective based on the visual pyramid is by Leon Battista Alberti in his treatise *On Painting*, written in 1435,[3] slightly before the Gutenberg invention. This lag is useful because it allows us to underscore that technology does not necessarily play a decisive role. It is not a question of reducing other human activities to technology or "explaining" them in terms of technology. Once again, I would like to refer to the analogy of the communicating vessels, a model that permits us to accord equal dignity to each domain of inquiry. I hasten to add, however, that in these cases the larger vessels undoubtedly have a greater responsibility and impact, so much so that they assume an almost eponymous role in the various cultures and establish an important point of reference, even if the "operators" in more sophisticated and avant-garde fields sometimes manage to arrive at great discoveries before others do.

Alberti describes with great clarity and rigour the concept of the pyramid produced by visual rays, which he describes as very fine threads that connect the object to the eye "where the sense of sight has its seat." Naturally, to produce the pyramid, the eye has to be reduced to a cusp, which is to say contracted to the size of a point. This is the origin of the notion of "point of view,"[4] which violates to an extent the physiology and even more the psychology of perception because, as we know, vision is binocular and involves apertures (the pupil with its adjustable size) that are anything but punctiform. Furthermore, what appears almost immediately is the related necessity of arresting time and movement. It is not enough for the visual source to be punctiform; it must also be immobile. Alberti also offers a solution for

3 I cite from the critical edition of the Italian version edited by Luigi Mallé (Florence: Sansoni, 1950).

4 The formulation and popularization of the "point of view" is the subject of a study by Marshall McLuhan in collaboration with Harley Parker, *Through the Vanishing Point: Space in Poetry and Painting* (New York: Harper and Row, 1968). Vasari alludes to a direct relationship between Alberti and Gutenberg, see Gian Luca Tusini, *La teoria prospettica di Leon Battista Alberti e le sue corrispondenze omologiche* [Leon Battista Alberti's theory of perspective and its homologies], in *Studi di Estetica*, 18 (1998).

binocularity, which is subsequently taken up by schools and academies of design influenced by Renaissance principles: the usefulness of closing one eye in order to neutralize the difference between one eye and the other.

The pyramid, the beam of rays stretching from the point of view to the surface of the object (divided into external and internal, with the central ray, the one that is orthogonal relative to the object itself, being privileged), constitutes a natural and physiological reality. Shifting into the artificial realm, Alberti explains that an "intersection" is required. A rigorously quadrangular plane may be lowered so as to cut the pyramid between the point of view and the object struck by the "rays," thereby capturing an image of the object, a *simulacrum*, on its surface. That plane itself can be real and not merely virtual. Alberti advises the artist to use a transparent veil that allows the rays to pass through. Better still if the veil is divided into squares forming a grid, in other words a system of horizontal and vertical lines. Thus the concept of the measurability of space is already at work, as it would later be when applied to the Cartesian axes.

If the veil is a physical reality and using it corresponds to an operational technique, Alberti is quick to identify an equivalent in nature that is more contemplative and can be found only in the visual field. In fact, it functions like an "open window," thereby confirming the intersection through a square. As is well known, the concept of the "window" that opens onto reality would become the *bête noire* of all the symbolic forms of space in the contemporary age, precisely because it is based on a reductive interpretation of perception (punctiform, static, instantaneous). Alberti, however, is aware that he is placing a straightjacket on artists, restricting their freedom of movement and obliging them to be nothing but an eye. Evidently, for him art is synonymous with technique and artists have to assume responsibility for the effectiveness and the technical-scientific functionality of their projects, and so they have to transform themselves into "photographers" *ante litteram*.

With respect to photography, however, the process identified by Alberti occurs entirely in the space outside the eye and so it corresponds more than anything else to a specular or reflective phenomenon. The "intersection" is essentially like the convergence of visual rays in a mirror. To produce the process on which photography is based, the pyramid would have to be transferred inside the eye in order to show that the pyramid occupies the space between the pupil and the retina. As occurs in the *camera obscura*, rays pass through an aperture and strike the internal surface opposite the point of entry, where they produce a virtual and inverted image of the object. Here, we have another arbitrary violence inflicted on the physiology of the eye; the surface of the retina is spheroidal while the surface of the *camera obscura* obeys the laws of plane geometry. Apart from this difference in the conception of the pyramid (external and real versus internal and virtual or inverted), what we have, in any case, is a continuous process that stretches from Alberti to photography in the nineteenth century, a process that in turn constitutes a pillar of modernity.

Panofsky's masterful analysis is aimed at identifying the possible theoretical consequences of the invention of perspective at the level of the encyclopedia of sciences. He demonstrates that a conception founded on the rigorous homogeneity of space emerges, consisting of points that extend in each of the three orientations (isotropy). This space is now quantitative and no longer qualitative, and the coordinates are up and down, left and right, in focus and out of focus. Zones glimpsed out of the corner of the eye are no longer valid. Every point is now determined by three objective measurements recorded on three axes. From this point of view, I cannot overemphasize the importance of the grid that Alberti recommends be drawn on the veil, an early recourse to the system of Cartesian axes and analytical geometry even if for the time being the veil is technical or material in nature. Panofsky does not overlook the objection that plane geometry based on point, line, and surface was already practised in antiquity using Euclidian elements, but in that culture, they never succeeded in transferring plane geometry to the technical field and making it the pivot of a coherent and unified representational system.

As for McLuhan, we know very well that his task was to pursue more daring homologies between a mental or theoretical realm (or a practical one if it concerns objects that are always symbolic like drawings and paintings) and a technological machine: the printing press. What, then, are their possible functional correspondences? Guided by his insight, we can divide these relations into three types. The first are material correspondences. During the entire modern era, the printing press emphasizes as a "normal" process the diffusion of the printed page, which is a flat rectangular surface that, due to its economy, becomes the preferred medium. Analogous to the page, the painted canvas, *quadro* or square in Italian, becomes the preferred medium in the symbolic sector of the visual arts, to the point of becoming synonymous with art itself. Even today there is the widespread belief that to speak of painting means fundamentally to speak of canvases. Furthermore, movable type, which uses discrete, standardized elements (separate from one another, unlike the continuous, interconnected graphemes of handwriting), is homologous to the concept of the point as the unit of construction for the homogeneous and isotropic space of modernity. At the level of the visual imaginary, the characters on a printed page are placed within a typographical grid that in turn confirms the normality of the rectangular pattern.

But we can also observe that, up to this point, such elements do not deal with perspective per se, which is to say seeing from a distance, at best confirming a Euclidian conception of two-dimensional space. When we read a Gutenberg page, the third dimension comes into play due to the fact that the viewer must take a position at a certain distance, which is effectively perspectival, and must cover the page with a beam of rays trying as much as possible to neutralize the binocularity of vision, forcing it to work in monocular or punctiform mode; in other words, the activity of reading the printed page promotes the establishment of the point of view. It is a process that can impose from below and with slow subliminal action what would otherwise remain a "high" theoretical construction of knowledge and, as such, is subject to being rejected by the average person or, at least, to not becoming an integral or stable part of the habits of that person.

AN INVARIANT OF THE MODERN AGE

Once Renaissance perspective is established, with all its homologies both at the level of material culture and the level of symbolic forms of knowledge, which I have tried to identify following Panofsky and McLuhan, what remains for me to do is to describe the extraordinary continuity and persistence of that perspective. We have to wait for the arrival of the revolutionary work of Cézanne to see that practice abandoned, which means that it enjoyed four centuries of relevance. If the technomorphic hypothesis formulated above is valid, the conclusion to be drawn is that, for a comparable period of time, there are no sensational technological revolutions capable of producing qualitative turning points, as opposed to inevitable adjustments and advances in the quantitative improvement of efficiency. In order to lessen the impact of imagining such a long period with no real change, I need to explain that the Gutenberg technology can be divided into two phases, as McLuhan suggests. In the first phase, the leading role (or eponym or indicator of cultural level) is played by a technology that operates in the field of communications and the transmission of knowledge (the book) and not in that of production (which remains artisanal) and locomotion (which remains reliant on animal power). This is followed by a second phase in which the technological revolution of machines powered by thermal energy (steam, then the internal combustion engine) dominates. How can we explain the fact that one of the most spectacular revolutions in the entire history of humankind did not produce an equivalent structure in the field of visual symbolic forms? In other words, why did it not bring about substantial changes in the inverted pyramid concept? Once again, it is useful to refer to a corollary of McLuhan's theory, according to which these two technologies – one based on information or communication (the printing press) and the other on production and locomotion (the steam engine) – are homological to each other. Basically, the same rational schemas, the same Euclidian geometric forms, are involved in both technologies. This is because they converge at the "high" or symbolic level of culture. During the humanistic Renaissance period, "modern" people had already proposed certain perceptual schemas, which also proved to be congenial to the mechanical phase of the first Industrial Revolution.

Curiously, this statement privileges the technological factor, but it also reveals its limitation in terms of the heuristic capacity of technology. The fact that technology leaves its mark for four centuries gives evidence of its centrality, but that same importance also neutralizes and turns it into an invariant. Due to its persistence, the technological factor loses incisiveness and explanatory power. This represents an advantageous correction to the presumed omnipotence of the cultural or material stratum and reduces the risk of determinism. The homologies between technology and "high" culture exist when they respect the logic of great epochal moments and they certainly do not represent a series of minor coincidences, almost reflex responses. In other words, it is not a matter of finding equivalents between fields at every stage of development, in every decade, and in a sort of mechanical way.

How, then, do we "enliven" this vast historical period that stretches across four centuries of western history and is among the most creative? It appears that factors in the other two planes come into play. These are the correlations among the various disciplines of knowledge in a given period, as well as the internal changes occurring within the field of the arts (which in turn are subdivided into the visual arts, literature, and the performing arts). In short, if the technomorphic concept plays the role of an invariant and, therefore, disappears from view (at least until the technological revolution of the contemporary age reactivates it in spectacular fashion) the variables of the other two levels acquire full force.

Stated differently, longevity and relative technical uniformity do not prevent the convention of perspective from being appropriated and articulated in very different ways, whether this diversity is consistent with ever-changing philosophical contexts or is a response to the need for change strictly within the field of the arts. There are different ways of feeling and treating perspective, depending on whether we live in the shadow of Platonic and Aristotelian thought or that of the new science; whether we believe in the primacy of rhetoric and the humanistic disciplines or deem these disciplines futile and favour an analytical approach. Things change as well depending on whether we arrive at analytical reason and the predominance of the physical-mathematical sciences via rational, deductive philosophy or via experimental, empiricist, inductive logic, whether in the interests of idealist historicism or those of naturalist Positivism. Up to this point, it is clear that I am speaking of great interdisciplinary convergences that connect developments in the visual arts with those in adjacent sectors; but there is no shortage of variants and fluctuations caused by purely internal factors.

What is more, McLuhan frequently makes us reflect on the lags between the effective date a new technology is introduced and the point at which it is assimilated profoundly at the level of common psychology and sensibility. A lag of this type can be noted with respect to Gutenberg's invention and its immediate homology (and indeed its prior appearance) in geometric and perceptual terms, namely, perspective – an immediate correspondence that perhaps can be explained by the fact that both inventions are technical in nature and, therefore, are especially close to each other. It would take several decades, however, before "typographic man" is born, that is to say human beings who have internalized the "high" culture consequences of this revolution to the point of constructing for themselves an appropriate logic and psychology. At this point, we need to recognize an important lag. The printing press, with its analytical and mechanical potential, takes its first steps when, conversely, Humanism celebrates its glories, proclaiming the rebirth of rhetoric, eloquence, and knowledge based on a proudly anthropocentric image of humankind.[5] Perspective cannot help but reflect this, which can also be said of how the first generations of the 1400s make use of it. I am referring to the contemporaries of Alberti, for example, Masaccio, Beato Angelico, Paolo Uccello, Filippo Lippi, and Domenico Veneziano (all born around 1400). Perspective is established at the mental level, but it is not followed up on in terms of its sensorial consequences. These would call for the human protagonist to be dissolved into the environment, into the air (Alberti already notes that light rays weaken as they recede from the object because they must travel through the atmosphere). The representation of depth is handled skillfully by theoreticians proud of their recently acquired mathematical knowledge, but not to the extent that the effect invalidates or weakens the power of the composition and especially the display of the human actor, who goes on being the master of the cosmos. More than ever human beings are seen as masters precisely because they possess the magical principle that allows them to subordinate all the elements of creation to the rigid control of their point of view. The situation does not change for the next two generations. The second includes artists born around 1420–1430 (Piero della Francesca, Andrea Del Castagno, Antonio del Pollaiuolo, Andrea del Verrocchio, Giovanni Bellini, Antonello da Messina, Carlo Crivelli, Andrea Mantegna, and Cosimo Tura). Among other things, how could we ignore the extraordinary concentration of dates of birth for these great artists, which is around 1430? The other generation is comprised of those born between 1440 and 1460 (Sandro Botticelli, Luca Signorelli, Domenico Ghirlandaio, Pietro Perugino, and Bernardino di Bello, also known as Pinturicchio). This fact allows me to reiterate that not even a generation is an indispensable unit of time to account for variation. In other words, a generation does not necessarily have to break with the preceding one. Sometimes, even frequently, its task can be to continue, expand, or strengthen the achievements of the preceding generation or to normalize these achievements, developing them fully in terms of quantitative criteria. Perhaps this is the historical task of the generation of Botticelli and Signorelli who came to emphasize beyond all limits the linear and anatomical rigour already present in the works of their pre-

5 For this, I refer the reader once more to my *Retorica* [Rhetoric] (Bologna: Fausto Lupetti, 2011).

decessors and, therefore, erase from within the tendencies inherent in the concept of perspective.

GIORGIO VASARI AND HIS THREE *MANIERE*

Not infrequently, from within the generation whose task appears to be that of "normalizing" and emphasizing quantitatively the accomplishments of the previous generation, an innovator emerges, someone who turns the page, even though he or she is a contemporary of those whom he or she challenges. Cézanne, a generational member of the Impressionists performed such a role and undoubtedly, for the epoch I am presently examining, the same role fell to Leonardo da Vinci, born in 1452. Da Vinci takes the first important step toward aligning practice with the spirit of perspective. For him, the primacy of the outline, "circumscription" for Alberti, comes to an end, superseded by aerial perspective, which takes into account the atmosphere and gradually softens profiles (*sfumato*). There is a similar revolution in the homological correspondences among the disciplines. In fact, it marks the end of anthropocentrism, the pre-Galilean concept of the cosmos, and, at the level of the logic faculties, it marks the beginning of the decline of rhetoric (Leonardo liked to think of himself as "a man without letters"), to the advantage of analytical reason and the mathematical method. Nevertheless, alongside these consequences of an interdisciplinary nature, repercussions inherent in the internal history of styles or *maniere* are even more striking. Stated differently, art historians, who are victims of their own "specificity," have tended to privilege the latter over the former. The spectacular stylistic change created by the artists of the 1400s and by the proto-fifteenth century Leonardo was noted and even theorized at least twice by two scholars: Giorgio Vasari, who can be thought of as the major theorist of the phenomenology of styles and the first to have a clear understanding of it; and Heinrich Wölfflin, who is the most coherent and enthusiastic follower of Vasari's method. As is well known, in the famous *Proems* to his great work, *Le vite* [The lives],[6] Vasari identifies

6 Georgio Vasari, *Le vite de' più eccellenti architetti, pittori et scultori italiani* [The lives of the most excellent Italian painters, sculptors, and architects] (Florence: Sansoni, 1966), edited by Rosanna Bettarini with a comment by Paola Barocchi. For comment on Vasari's theories and in general on Renaissance treatises on art, see my contribution to *Letteratura italiana, storia e testi* [Italian literature: History and texts], vol. v (Bari: Laterza, 1974), 106–9.

three *maniere*, a term substituting "style," which did not exist in those days. The first *maniera*, which applies to Giotto and other masters of the 1300s, is not relevant here since it precedes the invention of perspective. The second corresponds to the three generations of fifteenth-century artists mentioned above and is characterized by a "hard" and analytical composition, by the display of anatomy and the immanence of the body. The third *maniera* is the "soft" style of the flesh, gracefulness, distance, and *sfumato*. No one surpassed Leonard as master and exemplar of this *maniera*, especially when he urged his followers not to paint hair individually, but as a soft mass and to soften the body rather than depict it rigidly, exhibiting excessive knowledge of external anatomy.

Many centuries later, Wölfflin[7] appeared with his binary system, which is one of the most important interpretive instruments in visual arts. According to this schema, the early 1400s are characterized by "closed forms," while from Leonardo onward we have forms that are relatively "open." Among other things, Wölfflin is credited with having described these two phases in terms of a general sensibility (i.e., aesthetic, both in its etymological and Kantian meaning). To produce closed forms means to adhere to the body, to subject it to a kind of tactile verification at close range, which is a feature of a pre-modern and pre-Gutenberg culture (an important confirmation of the fact, as noted above, that the "spirit" of the Gutenberg innovation is not immediately established beyond its mere technical or instrumental application). Leonardo's revolution, instead, signals the triumph of the visual or the artist's relationship with distance and the full simulation of depth. In other words, he takes important steps toward a modern introjected and assimilated spirit. The "open window" concept approaches full realization in his work (which also explains why, conversely, he is thought to be rather remote in terms of the contemporary sensibility, as that same sensibility often rejects the age of Raphael, preferring to find echoes and correspondences in pre-modern, pre-Raphaelite,[8] or even primitive art that antedates the invention of Renaissance perspective).

7 See Chapter 6.

8 As we know, the great theorist as regards this accusation levelled against the age of Raphael was John Ruskin, followed by William Morris. Those who practised it in art were the Pre-Raphaelites, for which I refer the reader to my study bearing the homonymous title (Milan: Fratelli Fabbri, 1967), as well as the next chapter in the present volume.

But while Vasari predictably stops at the triumph of the *terza maniera* [third style], offering only the occasional criticism of the degeneration of the Mannerists, Wölfflin, who has a considerably more complex historical framework to control, develops further Vasari's schema and comes to the conclusion that the sixteenth century is only relatively "open" since it is still attached to the equilibrium of anthropocentrism. He also concludes that the sixteenth century is followed by a seventeenth century that is "open" in a completely different way since it is cosmocentric; as such, it makes the classical Renaissance era appear to be a "closed" phase (that is, fixated with the search for symmetry, equilibrium, and centrality).

Though a considerable distance separates them, there is a certain similarity in the roles played by Leonardo and Cézanne. Both come to the fore in precarious times and find themselves inserted into a generational space where they do not belong. Cézanne is placed in the company of the Impressionists, artists who perfect the modern system of perspective adapting it to photographic response (thus, basically achieving the theoretical rigour advocated by Alberti, the founding father). Leonardo is placed among such contemporaries as Botticelli and Signorelli who pursued "closed" forms (Vasari's second *maniera*) and the excesses of humanistic design. Thus, his teachings are followed only by those who belong to the next generation, the great canonical figures, including Raphael, Giorgione, Titian, Fra Bartolomeo, Andrea del Sarto, and Sebastiano del Piombo (born in the last two decades of the century). Leonardo's teachings are only partially followed, however, since the time is not ripe for full acceptance of the revolutionary force of aerial perspective, with the attendant rejection of rhetoric and Humanism in favour of the analytical-mathematical method. His young followers are certainly aware of the usefulness of a certain amount of gracefulness, depth of field, and atmospheric softening, but they also feel the need to defend rhetorical and compositional values. We know that their contemporary, Michelangelo, the traditional rival of Leonardo, pushes the power of the display of human anatomy to the maximum, subordinating every other value to that aspect of the work; he almost creates an "open" treatment (macroscopic, flexible, extremely elastic, and artful) of the "closed" humanistic universe typical of the preceding *maniera*, that of the 1400s. In short, we have the ambiguity of this generation of great artists suspended between the ancient and the modern, between tactile values and the triumph of the visual, between display of anatomy and mastery of the backdrop and landscape, between a fully realized modernity and a residual archaism and rigidity.

This great generation of artists is followed a few years later (not the requisite 20 years, but in many cases slightly fewer) by a generation that appears to achieve a revolution or at least to express the desperate need for novelty while remaining respectful of some common traits: playing the card of complication and exaggeration. Here I am referring to the first generation of Mannerists (Pontormo, Giulio Romano, Parmigianino, Bronzino) followed by at least a second generation (Tintoretto, Jacopo Bassano, Pellegrino Tibaldi). The latter group is separated from the preceding generation by the requisite twenty years. The mannerist revolution (or distortion) reminds us that the course of history is definitely not linear; in fact, it was the historical task of this generation to arrest the move of the canonical artists toward a fully realized modernity and to return to the archaic features of early fifteenth-century art, thereby revitalizing rhetoric in its most refined and sophisticated features, almost a *trobar clos* (a closed form of poetic style). In other words, the dominance of visuality and fusion with the cosmos comes to a halt, while the tactile values of the immanent body temporarily reassert themselves. This is the first in a series of waves alternating between a response to the "progress" inherent in the Gutenberg technology and a return to an ancient, acoustic, tactile space that relates more to motor function than to visual and mathematical criteria, and is qualitative rather than quantitative. (Consider how this effectively leads to a revival of curved, spiral forms that resist modernity but are appropriate for a pre- or postmodern sensibility. This also explains the reasons for which the contemporary age has re-evaluated and appreciated the mannerist phenomenon).

Here too, there are exceptions or lags in the dates of birth. For example, Ludovico Castelvetro[9] is born in 1505 and is therefore a contemporary of the most daring Mannerists. In the context of literature, however, he is the most strenuous defender of Aristotelian poetics,

9 See the chapter dedicated to him in *Poetica e retorica* [Poetics and rhetoric] (Milan: Mursia, 1969) as well as page 106 of *Letteratura italiana, storia e testi*, [Italian literature: History and texts], vol. v (Bari: Laterza, 1974). As for Mannerism, see my *Maniera moderna e manierismo* [The modern *maniera* and Mannerism] (Milan: Feltrinelli, 2004).

calling for verisimilitude and naturalness (even though these are filtered through a rigorous rationalism), which amounts to a rejection of the excesses of rhetoric. He is essentially a great "modern," a great precursor of seventeenth-century Rationalism. He is also a supporter of the crucial event that is the Reformation, which occurs at the level of "high" or symbolic culture. The Reformation is one of the first phenomena where the Gutenberg logic is internalized at the level of ideas and mental attitudes (the end of a complacent, pompous, extroverted, and spectacular orality and a transition to introspection, as well as the appreciation of simplicity and nature).

Born about one generation later (therefore in synchrony with the second wave of Mannerism), Cardinal Gabriele Paleotti (1522–1597)[10] gives theoretical expression to the visual concepts of the equal but contrary phenomenon in relation to the Reformation, and that is the Counter-Reformation. Today, these two great, critical events in the life of religious forms appear increasingly similar and convergent in terms of their basic objectives. They collaborate in accelerating the advent of Gutenberg's modern "typographic man" who is disdainful of rhetorical magniloquence and prefers to replace respect for verisimilitude with the silence of vision and introspection.

We have to wait a bit longer for the generations of the second half of the sixteenth century to find the true protagonists of the anti-mannerist revolution. In this regard, recent studies[11] have restored the credit due to the three Carraccis (Ludovico and his cousins Agostino and Annibale, born between 1555 and 1560), who played the historical role of interrupting the involution or implosion of the Mannerists by resuming the study of the canonical artists of the modern style at the point where it had been interrupted, thus leading to clear and explicit "modern" results, which is to say, a full, open, extended, and coherent language. All this comes about despite the diversity of inten-

10 *Letteratura italiana, storia e testi* [Italian literature: History and texts], vol. v (Bari: Laterza, 1974).

11 In particular the *Biennale d'arte antica* [Biennial exhibit of ancient art] organized by the Ente bolognese manifestazioni artistiche, founded by Cesare Gnudi and devoted to Guido Reni in 1954, the Carracci family in 1956, the masters of seventeenth-century painting of the Emilia region in 1959, classical landscape in Italian painting of the 1600s in 1962, Guercino in 1968, Federico Barocci in 1975, and the art of the 1700s of Emilia-Romagna in 1980. A new series was devoted to artists up to Giuseppe Maria Crespi from the end of the 1980s to the early 1990s.

tions (as though there were an almost intentional distribution of roles among them), by which Ludovico develops the middle style corresponding to Cardinal Paleotti's Counter-Reformation precepts of a discourse acceptable to the masses, while Annibale pursues a high style and a heroic language that is already literally neoclassical and, as such, suited to an audience of princes and noblemen. The moderate naturalism of Ludovico is subsequently reiterated, reinforced, and dramatized by Caravaggio (who was born in 1573, almost a generation later).

THE SEVENTEENTH CENTURY AND ITS "SOULS"

We now enter the very complicated and stormy 1600s. I would like to make it clear that I do not pretend to solve the many intricate problems of that period. I simply suggest a series of possible operations[12] that indicate how culturological instruments, in their various forms, can be applied (how they can be applied in the future in a much more extended and systematic fashion than is the case in this particular work).

The valuable "internal" stylistic category that emerges from the morphological reading suggested by Wölfflin applies generally to the entire century and especially to the first half of the 1600s. According to this paradigm, the century is a long phase of "open forms" where the golden equilibrium, centrality, and symmetry of the Renaissance come to an end. Their place is taken by respect for "distance," for the values of the background, the presence of atmospheric masses, with the subsequent development of *chiaroscuro*, the picturesque, and the dialectical dynamic that pre-empts the emergence of individual features. This is a general characterization that appears to be stronger than the numerous internal distinctions emphasized previously. We need to distinguish, however, between the naturalist trend (Ludovico Carracci's moderate version and Caravaggio's exaggerated version) and the apparently contrary trend of classical landscape (from Annibale Carracci to Nicolas Poussin and Claude Lorrain), without overlooking the fact that, between these two divergent styles, there is a line

12 I have performed these operations to an extent in my book *Arte e cultura materiale in Occidente. Dall'arcaismo Greco alle avanguardie storiche* [Art and material culture in the West, from the Greek archaic to the historical avant-gardes] (Turin: Bollati Boringhieri, 2011).

of development that includes Peter Paul Rubens, Gian Lorenzo Bernini, Francesco Borromini, and Pietro da Cortona, as well as the theoreticians of *concettismo* [conceptism] in literature, for example, Emanuele Tesauro, Matteo Pellegrini, Sforza Pallavicino,[13] and Baltasar Gracián, all born in the years straddling the two centuries, which is to say the last years of the 1500s and the first years of the 1600s.

In this regard, as it is well known, we have the intriguing historiographic problem of the Baroque,[14] to which an elastic meaning can be applied, to the point of making it coextensive with the three broad trends indicated above. We can also consider a more restricted meaning, where the term corresponds only to the last of the three trends; this seems to be consistent with the perception that contemporaries of the phenomenon would have had. Speaking figuratively, we could also say that it is a question of wave mechanics. That is, do we want to use the metaphor of long, macroscopic waves or that of short, choppy waves? I have already said that Wölfflin's long wave, whereby the entire century is characterized by "open" forms, can be confirmed from the standpoint of style. If the term corresponds to the cluster of characteristics identified by the Swiss critic and re-examined by me in the present study, Rubens and Caravaggio, Rembrandt and Poussin, Gaulli and Ruysdael would be "open" as well. This long wave, however, can contain three smaller waves synchronized with one another, which would be the three styles that art historians usually mention: Naturalism, the Baroque in the restricted sense, and Classicism. These smaller waves, however, could be enlivened by a play of oscillations within the long wave, oscillations that break up the wave into a series of "closed" and "open" segments that alternate between the ancient and the modern, the neo-rhetorical and the anti-rhetorical. The specific implications of the always meaningful dates of birth tend to support the synchronic interpretation (of smaller waves carried by the

13 For Italian theorists of "conceptism," see my contribution in *Poetica e retorica* [Poetics and rhetoric] (Milan: Mursia, 1969). See also Alberto Asor Rosa, *Letteratura italiana, storia e testi* [Italian literature: History and texts], vol. v, part I (Bari: Laterza, 1974), 64–7.

14 See Enrico Castelli's *Retorica e barocco* [Rhetoric and Baroque] (Rome: Bocca, 1955), Luciano Anceschi's *Le poetiche del barocco letterario in Europa* [The poetics of the literary Baroque in Europe] in *Momenti e problemi di storia dell'estetica* [Periods and problems in the history of aesthetics], vol. i (Milan: Marzorati, 1959), and Giulio Carlo Argan's *Storia dell'arte italiana* [History of Italian art], vol. iii (Florence: Sansoni, 1968).

large wave, without the need for a dynamic juxtaposition of the two). From these dates emerge many unusual, forced synchronies, contrary to the different and sometimes antithetical results emerging from the choice of poetics and style. Galileo Galilei and Giambattista Marino are virtual contemporaries (born in 1564 and 1569, respectively). Francis Bacon is born slightly earlier, in 1561. René Descartes, born in 1596, is a contemporary of true Baroque artists, such as Bernini. In each case, the cultural "operator" was generally receptive to the idea of an "open" universe that is cosmocentric and rich in potential. In this sense, the validity of Wölfflin's categories is confirmed; however, both at the level of style and at the level of interdisciplinary homologies, the works proposed by each of these "operators" as a response to a universe commonly perceived as "open" and infinite are very different. Galileo and Descartes see their respective teachings conflated as a common reaction to rhetoric and as privilege accorded to analytical-mathematical and experimental reason. The responses to this in art are the stylistic features that emerge in the form of a classicizing taste or archaism (in a sense, therefore, "closed" relative to bolder Baroque alternatives).

Instead, from Marino to Gracián and Tesauro there is a current that relies heavily on the tools of rhetoric and dialectics in an effort to redouble ingenuity and virtuosity with respect to *concettismo* in order to keep the space between them from opening up like a terrifying chasm. In short, for the first group, i.e., the rationalists and classicists, the infinite is to be confronted courageously with the silent instrument of vision and numbers, while for the second group, it is better to enliven that infinity with the eloquent sound of speech, of excited and noisy declamation. These are different responses to the same epochal moment, which can be seen as synchronous waves or appear in diachronic succession since in this specific realm of the visual arts, the Baroque phase (in the restricted sense), flares up in the first half of the century, while the second half of the century is characterized by a lively reaction against the first in the name of rationalist Classicism. (However, we cannot forget that this was already preannounced by Annibale Carracci).

At this point in our study, we have to remember that an important variable is neutralized, as noted above, that variable being the technological factor, with its equivalent in perspective. Technology remains neutralized for more than two centuries, bringing about a real reduction in the potential for innovation and variation. It had never occurred in the history of western culture, and perhaps will never occur again, that for centuries the visual, symbolic forms were not called upon to create their original conceptions of space. For example, in the relatively short period of the contemporary age (barely two centuries), we can say that every generation, school, and tendency has produced its own model of space. During the long modern period, instead, Renaissance perspective constituted a great invariant, a codified "language," even though it was possible to pronounce or speak that language with different accents and tones. As a result, since the spatial model could not vary, artists emphasized certain "internal" aspects of the model. This is why the dichotomies classical-anticlassical, rational-baroque, and closed-open appear repeatedly or at alternating intervals. There are many "returns" or revivals and so, if we wanted to characterize the 100 years from the second half of the 1600s to the first half of the 1700s, we would have to multiply the number of classicizing waves and revivals of the Baroque (Italian *barocchetto* or late Baroque and Rococo): alternations of low or naturalist style and elevated style, of colour that is earthy, tawny, and one that is silvery, and so forth.[15] This is accompanied by a progressive fading of reflected light, a muddying of the waters, confusion of vocabulary, and a clouding of the issues, followed by disorder and a growing din, which push people to desire a counter-measure, the advent of a new order, almost a new beginning. In short, at a certain point the modern age feels the need to start over again without, however, being able to knock down a crucial barrier since the basic conditions of material culture have not changed. The printing press remains the basic instrument of communication. At most, it is supplemented by the synergetic effect of the machine-based Industrial Revolution, but only in order to demand greater rigour and not to impose radically different styles.

15 The *Biennale d'arte antica* of 1980, dedicated to eighteenth-century Emilian art brought to light these two styles represented by Giuseppe Maria Crespi and Donato Creti, respectively (styles that came about after the bifurcation that developed in the time of Ludovico and Annibale Carracci or that of Guido Reni and Guercino).

5

The Birth and Evolution of the Contemporary Age

AN ASYNCHRONY BETWEEN SCIENCE AND TECHNOLOGY

I will have to apply the same criteria with which I defined the modern age to what the handbooks refer to as the contemporary age. It would be useful to start by recognizing the appropriateness of such a distinction; otherwise, the years stretching from the Renaissance to today would seem too long and undifferentiated. On the other hand, if the term "modern" were applied only to the last two centuries of that same timespan (the 1800s and 1900s), or simply to the 1900s as common sense and common usage of the term would suggest, what term would we use to describe the centuries that precede this period? Since there is a great deal of confusion with respect to terminology, in the sense that modern and contemporary are semantically equivalent, I have preferred for some time to replace the weak term "contemporary" with the more specific "postmodern." This is because the word has an immediate intrinsic advantage in the very way it is composed; the prefix "post" determines the meaning in a clear fashion, suggesting as it does a chronology. As such, it refers to a great historical timespan that comes "after" modernity. No one could misuse or misunderstand the term, which would occur instead if we continued to use the weak and ambiguous "contemporary." From another point of view, "postmodern" is marred by a certain ambiguity given the wide and inconsistent use made of the term today, where it is used to speak about recent events.[1]

1 In this regard, see my *Il ciclo del postmoderno* [The postmodern cycle] (Milan: Feltrinelli, 1968). See also Annotated Bibliography, Chapter 5.

In the pages that follow, I refer to a contemporary or postmodern age, taking the adjectives to be synonyms. As for periodization, once again I have to rely on the most traditional handbooks. In fact, I find the date 1789 generally acceptable as the boundary line between the modern and the contemporary age (the postmodern age in my own revised terminology). This time, however, we are not dealing with one of those surface events like the birth or death of great personalities, pageants celebrating illustrious dynastic families, or diplomatic and military incidents, which handbooks prefer to cite in other cases. The outbreak of the French Revolution appears to be an event that has a certain force and cogency, because it is tied to an important "set" of political and institutional circumstances: the collapse of the absolute monarchy and the feudal privileges associated with it, and the subsequent legal constitution of the State founded on representational democracy, which is the basis of a capitalist middle-class society.

But due to these intrinsic characteristics of the French Revolution and its consequences, we must ask ourselves if that "set" of circumstances does not, in fact, reinforce "modernity" rather than end it and set in motion a new set of circumstances. In other words, as far as French history is concerned, 1789 seems to coincide with the actualization of premises that already existed in the "modern" technological revolution (individualism, consisting of the ability to read the Bible and other sacred texts with one's own eyes, thanks to the advent of the book and the printing press, the Protestant Reformation that follows, and its "homological" links with the dawn of capitalism, and so forth). More than a "new" threshold, 1789 stands out as a phase that confirms and sanctions the advent of modern individualistic, capitalistic, bourgeois society invested with the task of driving off the ghosts of the ancient medieval world, ghosts already dispelled for some time by the various technological turning points and their economic and social effects. For some time, the economy was no longer based on land, agriculture, trading in kind, and short-term artisanship. Therefore, the political and institutional events of the French Revolution also appear not to be "foundational" but rather surface phenomena, like the aftershocks of an earthquake that occurs deep underground and some distance away.

Furthermore, I am bound to respect a method that has proven to be functional and so I need to identify a dominant technology that characterizes the contemporary or postmodern era and reshapes the entire cultural cycle to which that technology corresponds. For the modern age, I relied on McLuhan's general theory, according to which we have to think in terms of a Gutenberg Galaxy (placing all other responses in the various sectors of culture in a relationship that is "on par" with this innovation). Fortunately, the Canadian scholar offers us an equally authoritative suggestion for the contemporary age – a courageously unified and provocatively simple one. For McLuhan, our epoch relies on "electrotechnics" (the use of electric energy or, better still, electromagnetism as a whole) for carrying out practical tasks, a term that may also include electronics in its definition. The word coined for this domain, "technotronics," may be considered the typifying feature of the current phase, namely, the late twentieth century.

The symmetry of the method applied to the study of the modern age appears convincing; however, I need to point out a possible chronological discrepancy. There are two key moments in the modern age. The start of the "set" of circumstances from the standpoint of the history of technology (the invention of the printing press in 1450) and the start of these circumstances from the standpoint of casual epiphenomena (the death of Lorenzo the Magnificent and the discovery of the Americas in 1492) are very close together and almost coincident. This is not the case for the contemporary or postmodern age. In fact, if we remove the dates of political and social relevance pertaining to the French Revolution of 1789 and we look for more compelling dates relative to innovation in electrology, the latter offers us an abundance of discoveries, ideas, and debates of a strictly scientific nature, with little in the way of technological outcomes. In other words, in 1789, the electrical (electromagnetic and electronic) realm for the most part concerns physics and epistemology, sectors of culture we call "high" or symbolic, but it does not trickle down to the practical-material level where technology usually operates. In that environment, the most advanced sectors of western society are dominated by electromagnetism's counterpart: the mechanical-industrial revolution based on the exploitation of steam power or other sources of heat energy. Taking the year 1789 (or thereabouts) as the concluding phase can be justified in material terms as well. At that time, the "steam barons" come into existence; they are clearly modern, bourgeois capitalists who distance themselves from the agricultural and artisanal modes of production based on animal or human power. Operating at the "high" institutional level, that same class organizes for its own benefit a system of democratic, formal, representational freedoms (the famous universal principles of liberty, brotherhood, and equality promoted by the French Revolution).

At this stage, what do we do with my hypothesis of a fundamental technomorphism (that is, a culture in synchrony at all levels with the dominant technology of the period)? The safest thing to do is to wait until the technology of electromagnetism emerges fully, in other words, beyond the first theoretical or laboratory results, which are ineffectual in terms of practical consequences. In this case, we would need to wait almost a century, the 1860s, to find two important events related to electrotechnics. The first occurs in 1860, when Italian physicist Antonio Pacinotti (1841–1912) invents the dynamo named after him. It is essentially a device that uses the combined effects of electricity and magnetism to produce mechanical work, which causes an inert object to rotate. The era of electromechanics begins at that moment, and we can consider it to be a kind of happy compromise between the two levels. The driving role played by the "machine age" at the centre of the processes of production and locomotion does not diminish. Work, which is performed by the muscles of animals and humans in antiquity and during the Middle Ages is still carried out by inert mechanisms, but the energy that moves them is no longer heat (steam or the combustion of organic substances, such as coal and other hydrocarbons); it is "white" electrical energy. But it is only an intermediate surrogate, a certain phase in the chain. In the end, we are still dealing with functioning machines with their heavy and restrictive performance; as a consequence, it can be argued that we have not emerged yet from the "modern" world. This is even more the case since electrical energy can still be produced from heat sources by using hydrocarbons, as still occurs in many parts of the world. Turbines generate the "new" electrical energy, which causes other turbines to rotate in a kind of set of Chinese boxes or a hybrid modern-contemporary system of energy production.

The indicators of innovation seem even greater in another technological achievement from the same decade, namely, the laying of the first transoceanic telegraph cables from Europe to North America in 1866. From that moment forward, telegraphic communication connects the two continents in fractions of a second, since electromagnetic waves move at the highest speed possible in the physical world, almost 300,000 km per second (like light, a phenomenon that falls within the sphere of electromagnetism). The fact that such a velocity exists and that, although very high, it is a finite value that cannot be exceeded, is one of the most typical features of our culture and one of the highest achievements of contemporary or postmodern science, as formulated in Einstein's theory of relativity. In addition, this incredible velocity is first put to use at the technological level in telegraphic communication, which represents a giant qualitative leap forward with respect to the velocity attainable by even the fastest and most sophisticated mechanical devices. Consider the unbridgeable gap between the time it took a telegram to travel from London to New York and for a letter to be sent across the same distance by steamship in 1866. In the first instance, that would be fractions of a second (the time it took for the circuit closed by the telegrapher in London to engage the entire system). In the second instance, it would involve weeks of difficult travel at the mercy of weather conditions. As a result of the improvement in mechanical technologies, the steamship today is replaced by the jet, some of which can reach supersonic speeds. However, this implies an improvement in performance at the quantitative level, not at the qualitative level. From such a perspective, the gap between communication by electronics and communication by mechanical means remains immense and unbridgeable. It is worth pointing out that electromagnetism has undergone only quantitative not qualitative enhancements. In other words, given that the speed of electromagnetic waves is absolute, the message they carry (the waves) takes the same amount of time to travel between Europe and North America whether it is sent using an antiquated telegraph from a century ago or the most efficient product of current science. Also noteworthy is the fact that the telegraph is a forerunner of a family of applications that, from general electrotechnics, develops into a more specifically electronic family. The latter can be defined in terms of weak currents used in information and automation processes. In addition, the telegraph is fully adapted to such coordinates, even in its antiquated role. It is subsequently replaced by increasingly sophisticated devices ranging from the telephone to the wireless telegraph, which opens the way for radio, the cathode tube, and the age of television. But again, these are only technical improvements of the quantitative kind occurring within a qualitative revolution that is already implied in the telegraph. We can acknowledge the chronological discrepancy between 1789, indicated in textbooks as the start of the contemporary-postmodern age, and 1860 (or thereabouts) when we finally find a functional technology based on electromagnetism. Therefore, we could say that the contemporary or postmodern age begins only on this second date. Things would work very well even if we were to look for homologies in the other fields of culture, including the high-symbolic ones of the "arts and sciences," in which the present book is particularly involved. In fact, there is a fascinating correspondence, even in generational terms, between Pacinotti and Cézanne (1839–1906), who can be thought of as the artist credited with rejecting the "modern" perspective completely and experimenting with the use of curvilinear and spheroidal space, which is difficult to explain without resorting to homologies in the electromagnetic "field." In this regard, it is appropriate to include a third protagonist alongside the technological innovator Pacinotti and the great artist Cézanne. This figure, a great exponent of "high" scientific inquiry, is Scottish physicist James Clerk Maxwell (1831–1879), who is only slightly older than the other two but not completely outside the generational window. In the decade of Pacinotti's dynamo, the transoceanic cables, and young Cézanne's experiments with plastic spatial designs that are so odd and different from those produced by his contemporaries, Maxwell produces mathematical formulas and general conceptions that would be the basis of electromagnetism, even though he does not publish these insights until 1873, in *A Treatise on Electricity and Magnetism*. Nevertheless, historians of science agree in recognizing the innovative value of these contributions from which a "new" concept of the electromagnetic universe emerges, one that is finally capable of doing justice to the field rather than maintaining it within the constraints of the old parameters of the Newtonian mechanical universe. We no longer speak of discrete atomic entities that influence one another at a distance, like heavenly bodies in the gravitational field. The electromagnetic field is an organic structure that integrates, connects, and creates continuities rather than aggregating them. As a result, our contemporary culture must now rely on *Gestalt* models and think in terms of "wholes" or interactive matrices in all fields of inquiry.[2] This is precisely the visual turning point sought by Cézanne in that same decade, and we are really dealing with the formative moments of a cultural continent that can properly be called contemporary or postmodern since it has severed all ties with the preceding modern uni-

2 See Evandro Agazzi's excellent introduction of Maxwell's *A Treatise on Electricity and Magnetism* (Turin: UTET, 1973).

verse (perspectival, Gutenbergian, founded on atomic aggregations and discrete elements).

In sum, we can safely speak of a culture – art, science, and technology – that is contemporary or postmodern in terms of a timespan stretching from about 1860 to today. In the previous edition of this book (1982) and in many of my lectures and publications, I made this claim and I clearly do not regret doing so. I simply would like to mention, among my other contributions, the one with the specific title *Arte contemporanea* [Contemporary art], intended to be a fairly analytical history of what happened in the visual arts during this period of time. (Even more telling is the subtitle, *Da Cézanne alle ultime tendenze* [From Cézanne to recent trends]).[3]

But there is still a glaring gap that lasts almost a century between the year 1789, which textbooks insist on identifying as the onset of the contemporary age, and 1860, which is the beginning of a shorter contemporary period reliant on electrotechnics and emerging electronics. In theory, there is room in the intervening space for an extended modernity, which witnesses its fullest and most typical phases owing to the development of the mechanical-industrial revolution. Indeed, we have the paradox of a culture that undergoes its most intense phase almost *post mortem*, that is to say, after it has been proclaimed to be surpassed. On this matter, however, the textbooks could claim to be completely logical in selecting 1789. They were intended to signal the beginning of a process destined to see the bourgeois-capitalist class achieve its greatest success when, in fact, that class wanted the freedom to confront the steepest and most intense phase of its social and economic ascent. In short, we needed to enter the age of industrialism and to witness the triumph of machines powered by heat energy, after the links with the old regime were severed and the sacred canons of free enterprise, respect for private property, and the inviolability of the individual were established.

In a sense, everything suggests that 1789 should be discarded because it is too remote and not appropriate for the flowering of the electronic era, which comes later and more successfully. In the area of inquiry related to electromagnetic phenomena, however, there is al-

3 *Arte contemporanea* [Contemporary art] (Milan: Feltrinelli, 1984 and 2005).

ready some ferment toward the end of the 1700s. How can we overlook the important contributions of Italian scientists like Luigi Galvani and Alessandro Volta,[4] among others? In essence, we already have the first signals of the crisis of the "modern" edifice erected by Galilean-Newtonian physics, which developed into mechanical physics and represented a privileged and authoritative paradigm. The challenges to such an edifice increase because, among other reasons, the new frontiers in electrodynamics appear to contrast absolutely those parameters. On the other side of the coin, it is true that, for the time being, the uncertain and experimental nature of this inquiry does not succeed in producing concrete technological results. There is, however, a series of eloquent, fascinating homologies to be discovered in the other areas of culture. The vessels communicate and a new liquid circulates in a system of connected channels.

WILLIAM BLAKE, A GREAT INITIATOR

A great figure in the field of the arts on which the energies discussed above appear to converge is Englishman William Blake (1757–1827),[5] who plays the same pioneering role that I have attributed to Paul Cézanne, only a century earlier, and with the advantage that Blake had a greater number of interests. These are artistic in the strict sense (Blake was a draftsman and illuminator, partly in order to survive); he was poetic (his collections *Songs of Innocence* and *Songs of Experience* ensure him an important place in the Anglo-Saxon Parnassus), philosophical, religious, and moral (thanks to his prophetic books). But perhaps the attempt by some critics to identify different genres in his production where they come close to applying the concept of "modern" analysis is inappropriate and a bit pedantic since Blake was really the preacher, the prophet of the wave of energy in human beings about to appear in all expressive literary and graphic forms. His re-

4 Luigi Galvani, *Opere scelte* [Selected works], edited by Gustavo Barbensi (Turin: UTET, 1967); Alessandro Volta, *Opere scelte* [Selected works], edited by Mario Gliozzo (Turin: UTET, 1967).

5 See my "William Blake alle origini dell'età tecnetronica" [William Blake and the origin of the technotronic age], in *Rivista di Estetica* 32 (Turin: Rosenberg & Sellier, 1989). As stated in the Introduction to the present work, much of this material appears in *L'alba del contemporaneo* [The dawn of the contemporary age] (Milan: Feltrinelli, 1996 and 2008).

course to the written word allows us to make explicit use of his work in the present theoretical study, which necessarily privileges the vehicle of discourse. In *The Marriage of Heaven and Hell* (1794), perhaps the densest and most direct of Blake's prophetic books, written in the appropriate years, i.e., the 1790s, when the discoveries of nascent electrology appear with greater frequency, we can find many valuable statements, such as the following: "Energy is the only life and reason is the bound or outward circumference of energy"; "energy is eternal to light"; "those who restrain desire do so because theirs is weak enough to be restrained."

We could conveniently interpret all this in the context of the umpteenth revival of Platonism, as a reaction to the domination of rationalist and empiricist thought in the entire eighteenth century. In effect, a return to Plato and his disciple Plotinus has historically represented a return of the force of emotional, passionate waves, those described with the excessively simple and imprecise term "irrational." But how do we define this return if we do not want to reduce it to a cyclical, almost atemporal phenomenon? How do we fail to see an analogous pseudo-Platonism percolating in the thoughts and art of many other contemporaries of Blake, some of whom will soon be mentioned in this study? And how do we avoid the temptation of building a bridge (of hypothesizing a homology) between this upsurge of affective energies in the field of literature and art, on the one hand, and the real attention that the most advanced scientists of the same years dedicate to energies that are equally unrefined, those of animal electricity, atmospheric electricity, or conducting bodies, on the other hand? In both cases, a *quid* appears, disrupting the orderly grids of Cartesian and Newtonian thinking, forcing us to start over again or perhaps to contemplate a model of coexistence that is not very peaceful, where the old instruments of Reason are called upon to dampen enthusiasm, chop things into elementary particles, and measure objects with mathematical criteria. As expressed in one of the first aphorisms cited above, Blake sees Reason as a limiting principle, as a non-conductor. As a visual artist, he does not hesitate to give Reason a human face, i.e., that of an old man with a white beard, very authoritative and imperious with its flashing peaks. Blake likes to rename the creatures of an Olympus redesigned according to his personal vision. Using words that he coins, he proceeds to superimpose and intertwine multiple meanings in order to produce enigmatic words not easily deciphered. Thus, he christens that powerful, irate, despotic old man with the densest and most hermetic of terms, *Urizen*. According to some, this is an ironic, desecrating version of *your reason,* as if to say, "See what your reason is reduced to." This is the same reason on which the whole of modernity and the enlightenment age erected its successes: a force that freezes the vital forces, subjecting them to tight control. According to others, it could be a reference to the Greek root word *orizein*, which gives us *horizon*. Furthermore, the two etymologies, which are closely related phonetically, whereby reason sounds like *rison,* can be easily confused since such a conceptualization leads us to see the principle of reason (the intellect) as something that limits, constricts, and closes; in other words, there is a philosophy of the finite completely opposite to the one of infinity (let us recall the hedge in Leopardi's poem "L'infinito" [Infinity]). The first is a feature of the Age of Enlightenment, of "progress," in Leopardi's words "le magnifiche sorti e progressive" [the magnificent progressive destinies]. The second typifies an age felt to be true and authentic, but whose development is uncertain. Stated another way, Blake is prepared to swear on the value of his times with his indestructible prophetic faith, while Leopardi, born in 1798 and therefore a full generation later, senses that modern "reason" has come out of the fray not at all defeated and is about to resume its course in full force. After the eighteenth-century Enlightenment there would be another positivist-naturalist stage through much of the nineteenth century. With respect to these two non-conducting strata, the gushing of electrical, emotional energies announced by Blake is threatened and forced to recede into a mystical past.

In the meantime, however, Blake is not alone in sensing the advent of that energy flux capable of breaking through the limits, the "horizons," of old Rationalism-Empiricism. In terms of the literature, for example, he can be placed alongside an author who, in many respects, is quite different albeit much more well known, Wolfgang Goethe (born a mere eight years before Blake, in 1749). How do we explain that unexpected emergence of deep, uncontrollable, almost *a priori* passions, as represented by Werther in *The Sorrows of Young Werther* or Edward and Charlotte,[6] the "impossible" lovers of *Elective Affinities*? In each case, love goes beyond conventions and stands for an "affinity" or

6 Johann Wolfgang von Goethe, *The Complete Works*, 10 vols. (New York: P.F. Collier, 1903).

irresistible attraction between people who are already married and, therefore, should be compelled to resist the impulse. Resistance is possible in a system of mechanical physics or of inert, corporeal bodies, but not in the emerging field of electrical attraction that no obstacle can resist. This is also because the new reality comes to light: it is we who exercise our will, even though at a deep level of consciousness. Prior to this, the will was placed at the same level as rational-intellectual, conscious acts performed in a state of complete self-control, with eyes wide open in the waking state. From this point on (when Blake and Goethe begin to explore the new sentimental map), "we" are split in two: a superficial will controlled by *Urizen*, where we try to respect all the formal obligations, the norms of respect and social convention, and a "deep," unlimited, and uncontrollable will that constitutes a unique reality with the very wellsprings of Life and Energy. However, "we" are always ourselves. The personality is split into two halves that have different origins: one conforms to the outmoded criteria of the rationalist age (mechanical, Gutenbergian, "modern"), while the other adapts to the "newest" demands of the electric world.

With respect to the philosophical debate, these issues come together in Arthur Schopenhauer's (1788–1860) masterpiece *The World as Will and Representation*, a title that identifies immediately the two halves. One is the domain of "representation," which is aimed at applying the "modern" coordinates of space and time (it too falling within the mathematical and analytical parameters) corresponding to Renaissance perspective and the Cartesian axes; the other is the much more violent and pressing domain of *Wille zu Leben*, vitalism, or the immeasurable flow of energy, in other words, the new physics of electromagnetism (which still awaits its definitive formulator Maxwell) as opposed to the old Galilean-Newtonian edifice.

All these intuitions anticipate the major achievements that, in the same psychological, epistemological, and metaphysical arena, would be produced nearly a century later by Henri Bergson and Sigmund Freud (cited as representatives of a whole host of similar thinkers). But as occurs with all preludes, the one we find in the works of Blake, Goethe, and Schopenhauer (not to mention their many contemporaries) appears to be relatively weak, tentative, and on the verge of digressions or episodes of involution; at the end of the 1700s electrology seems similarly tentative and timid in comparison to the solid results that would follow. All my efforts here are directed to corroborating the contention that, between the 1700s and 1800s, there was a well-organized foretaste of contemporaneity and postmodernity capable of sensing the essentials of the coming era, but one expressed in forms still inadequate and underdeveloped. In a sense, such a situation was aware of the stop that it would soon experience as a consequence of the return of the modern with full efficiency in the form of Positivism and Naturalism.

We can now return to Blake's many aphorisms contained in *The Marriage of Heaven and Hell*, particularly the section of the so-called "Proverbs of Hell." What is described here is a mind fearful of the censorship imposed by a "superior" and punitive principle, even if the terminology used is imaginative or "literary" and still far from the rigour and pertinence that would be shown by Freud and psychoanalysis. For example, we read, "Prudence is an old maid" whereas Freud would probably speak of the problems of excessive suppression of the libidinial impulses, resulting in neurosis. In a way, this is recognized by Blake himself in another proverb: "He who desires but does not act experiences pestilence." Elsewhere, we have the metaphor of water that is vital only if it overflows, reaching a fluid and dynamic state, whereas it becomes pernicious if it stagnates. In this metaphorical vein, one of Blake's most quoted turns of phrase involves an opposition between tigers and horses: "The tigers of wrath are wiser than the horses of instruction." Freud would apply a more technical and neutral terminology to these notions and would refer to the clash between the "pleasure principle" and the "reality principle." He would also establish the unavoidability of a perpetual clash between the two, as occurs in the case of the negative and positive poles of electromagnetism. What is more, Blake acknowledged that we cannot do without the paternal image of Urizen and his beard laden with vernal wisdom. The important thing is that, alongside that image, we find an important space for the image of the baby Jesus and Christ, the Son of God, depicted as a luminous figure radiating energy. For the figure of God-the-son, Blake is prepared to offer some of his enigmatic names: Orc and Los, which can be read as anagrams of the positive terminals of this unusual cosmos: Cor and Sol. The sun or the hot heart of the new frontiers of energy defeat the rigours of "old" modern reason and the clash is undoubtedly accompanied by an emotional wave, which conversely becomes more circumspect in the final and more mature version that Freudian psychoanalysis will provide for such a model. But it will be an extension, a technical advance outline of which has already been drawn. The same can be said of the dynamic binary opposition in Bergson's philosophy where *élan vital*, a fluid and undivided process, is opposed to the intellect that segments, breaks up, disjoins time, reducing it to the intellect's convenience. According to Bergson, there is a "bad" or improper time based on spatial units, which enjoy complete and reciprocal homogeneity. But on the contrary, there is a notion of time that is heterogeneous and incompatible with space, a realm of fluid, qualitative processes. However, Bergson never established the connection between his own metaphysical entity and the physical properties of the electromagnetic field. In this regard, neither he nor Freud went much beyond their precursors from a century earlier. Here again it is up to culturologists, who operate from a greater historical distance, to pierce the partition that separates different sectors of the same culture. This usually involves connecting the "high" stratum of the arts and sciences with the "low" stratum of technology.

Following this incursion into the field of "ideas," I return to examining those aspects of the visual arts that are most relevant to the present study. Even on this front, Blake appears to be a great subversive, lucid and resolute in his aversion to the "modern" project. He condemns Newton at the epistemological level and rejects with comparable resoluteness those in the artistic sphere who are situated at the start of modern "representation," that is to say a verisimilar, naturalistic art that captures the atmosphere, the picturesque, spatial depth, darkness, and tangled and chaotic compositions, with the aim of suggesting the real caught in its spontaneous unfolding, as if looking through a keyhole. Blake's knowledge is approximate (in those days, a textbook on art had not been published) and he did not benefit from a trip to Italy, unlike his colleagues; but his instinct is unfailing, so he condemns Correggio and Titian who in effect are the best representatives of Vasari's *terza maniera* [third style] because they push Renaissance art to attain impressive degrees of freedom and illusionism. After them he criticizes those who revive tradition at the beginning of the 1600s by adopting the style usually defined with the broad label of Baroque – from Rubens and Rembrandt, right up to the exponents of this illustrious tradition that Blake saw blossoming in his own country, especially in Joshua Reynolds, whom he describes as the great corruptor.[7]

7 Indeed, Blake's most insightful observations on the historical development of European painting can be found in "Annotations to Sir Joshua Reynolds's *Discourses.*"

Among the "masters" of the Italian Renaissance, Blake spares Michelangelo and Raphael. In truth, Blake knows little about Raphael and there is no evidence that he was influenced by him in any tangible way. For the most part, he spares these figures because of their "celebrity" and we can infer that he was thinking primarily of the young Raphael, a pupil of Perugino, and therefore one who belongs, for Vasari, in the "second style," consisting of graceful lines, symmetrical and pure compositions that are still some distance from the features of the Vatican rooms where Raphael initiates a style of great naturalistic maturity, thereby creating a watershed separating an early from a later phase of his art. This is so much the case that Blake's successors would coin the term Pre-Raphaelitism, thereby signalling a much needed return to the appreciation of "primitive" art that precedes Raphael. But in some ways, Raphael is the first of these Pre-Raphaelites, at least if we consider the works he produced prior to his transmigration to Rome in 1508 and the start of his work on the Vatican rooms.

Blake's relationship with Michelangelo is very different. He expresses an unlimited and tireless admiration for him and repeatedly sings his praises to the four winds. We also find that prints based on the famous frescoes of the Sistine Chapel undeniably influenced Blake's art. In fact, the Englishman never visited the chapel since he was averse to the traditional trip to Italy, a trip he "cancelled" for many reasons, including economic ones undoubtedly, given the always precarious state of his finances. There were also ideal and theoretical reasons in that he was always a great traveller but in an interior landscape since he was disdainful of escaping into the physical world; at the end of the day he might have been disappointed by a personal examination of Michelangelo's frescoes on account of their chromatic opulence and swollen lines. Instead, line and draftsmanship were the primary attributes of the miniatures produced by engravers who spread Michelangelo's aesthetic message and invariably produced smaller versions, reduced to their graphic aspects, emphasizing them, and anticipating the additional reduction they would undergo in works that characterize their recovery by Blake, which will be described shortly.

Blake admires Michelangelo as a kindred spirit belonging to the same group of Neoplatonists who place the human figure at the centre of the universe without wasting time and energy on the marginal aspects of creation, as did the culture of Baroque naturalism. There is only the human being who occupies the centre of the composition but certainly not in a static or inert way. Indeed, what Blake likes in Michelangelo's work is the excess of energy with which the human figure is invested or the Titanism, meaning muscles swollen and distended from exertion. Let us remember that supporters of the official naturalist-baroque style of the 1600s and 1700s had turned their noses up on these same features, essentially relegating Buonarroti to a kind of limbo, not having the courage to challenge him but also not praising him – nothing like the boundless worship of Raphael, Correggio, and the great Venetian masters of colour. If anything, we need to hasten to recall that Michelangelo was the nearest and most important model for the Mannerists, that is, those waves of Renaissance artists that appeared at the dawn of the 1500s and the next twenty or thirty years. As noted in the preceding chapter, these figures are responsible for having pursued an anti-modernist project; they applied the brakes to the march of modernity as it was prefigured in Raphael, Giorgione, Titian, and Correggio (that is, as the institution of an art capable of simulating reality in the most compelling fashion). These anti-modernists liked to adopt archaizing forms, or at least artificial, sophisticated, and exaggerated forms. Michelangelo is the primary and most obvious source of this excess, to the point where he almost serves as a battle flag. It is possible to have a Michelangelo cult where one seeks to go beyond the golden mean, verisimilitude, and naturalness. His works resurface every time one wishes to practice an art that is artificial, so to speak, designed to increase the intervention by human beings, with their demiurgic and arbitrary character.

The Michelangelesque, and hence neo-mannerist, aspects of Blake's work are clear, and, through these, he aligns himself with those among his contemporaries who also subvert modernity. But what relationship is there between these stylistic features and the technological-scientific revolution based on electromagnetism, which I emphasize in order to claim that Blake belongs to the contemporary or postmodern age? More generally, what is the link between the technological features and the "return" to Wölfflin's categories put into practice by Blake as the Mannerists had done in their day? In fact, both revive the "closed" forms, as opposed to the "open" forms of post-Renaissance art (which is masterful in the simulation of spatial and atmospheric effects) and both emphasize surface, clarity, and a simple composition that is paratactic rather than dense or convoluted. How do we go from the formal stylistic sequence to the epistemological-technological one?

In general, the visual culture of modernity draws out bodies by projecting them onto a plane point by point, directing linear "rays" onto them and allowing these rays to return to the point of emission. Perspectival distance is an integral part of this visual concept insofar as the term *perspectiva* indicates seeing from a distance, crossing an empty medium, and seeing that does not disturb or present obstacles, whether physical or theoretical; conversely, it may also encounter obstacles since space is filled with physical particles, atmospheric gases, and clouds, but in this case, they too become objects that reflect visual rays. The mirror and the *camera obscura* are natural guarantees of the validity of a similar type of "reflection," of representing reality, to which both the theories of artists from Alberti's time onward and "machines," the technological inventions coming to photography and its derivatives, make reference.

But important differences appear in an electromagnetic system. First of all, the object is no longer struck by linear optical rays but by waves, which have such a high velocity that they make the presence of distance practically irrelevant, at least as far as our limited earthly domain is concerned. All this leads us to the effect of spatial flattening. In other words, the technotronic age no longer knows what to do with the ritual of accurately measuring distances on which, conversely, modern culture as a whole relied (perspectival, post-Renaissance culture). One of the features that allows us to determine if a visual image is "contemporary" (postmodern) or not, if what we are dealing with is a remote, pioneering form of the contemporary or a form closer to us, is precisely the reduction of depth, which can be exaggerated to the point of two-dimensionality and can emphasize surface texture. Postmodernity sets a course contrary to that of modernity in this regard; if anything, it builds a bridge of rediscovered solidarity with the representational canons of the pre-modern periods (which can also be called archaic or, using an imprecise but effective term, primitive). Thus, we have the origin of "primitive taste," which is almost always present in manifestations of postmodern culture.

In addition to flattening, the diagnostic-representational instrument, which consists in sending waves, lacks the technical capacity to record details; in other words, it does not concern itself with these details. Acoustic, hydraulic, visual, and electromagnetic waves round off their targets, bypass them, and return an outline of an image to the viewer. This leads us to claim that, since it tends to flatten, postmodern representation tends simultaneously to "abstract" and, as such, capture general schemas, types, essential structures rather than constellations of particulars. The images it produces are essential but not detailed. They do not deal with *de minimis* [minutia], like the proverbial praetor in the Roman juridical system. This is a dominant feature of the technotronic field, which is calibrated to grasp *Gestalts*, wholes, structures; it would be incongruous if it aimed to grasp particulars. For the latter objective, it seems better to use the old modern-perspectival paradigm, as well as its best technical descendant, photography.

If we return to the Michelangelesque elements in Blake, having identified the general characteristics of the postmodern condition, we can understand what pushed the English artist onto this path and what led him to make obvious distortions on the model he respected. What Blake likes in the work of the master of the Sistine Chapel is the athleticism of the human figures, which seem to mimic those we might find in an exercise manual where they strike predetermined poses. By comparison, Raphael's handling of the human figure in the same years appears to be subtler and freer, as he demonstrates in cycle of frescoes in the rooms adjacent to the Sistine Chapel. For his part, Blake certainly emphasizes the reduction of his figures to types or even to mannequins, making them undergo a reductive, regressive process that nudges them inexorably just short of the Renaissance-modern line, ankylosis, and anatomical contractions typical of "primitive" taste. Even though he begins with the great classical model, Michelangelo, Blake emulates the compact volumes of Giotto and Cimabue in a series of stages. This is also true because muscle contraction is accompanied by flattening, as though an iron had passed over the bodies, preventing them from swelling up violently and imprinting them onto the surface. It is as though Michelangelo's types or *maniere* [styles], which are always those of a great classic who possesses great freedom (otherwise we would not be able to understand Vasari's admiration for Michelangelo), were emulated by an impoverished, regressive copyist at that level of culture who was determined to produce a transcription of these figures in primitive-archaizing style. As a result, we can say that Blake "collapses" Michelangelo's forms, using the term in its astrophysical meaning, according to which there is a "collapse" when the stars of a galaxy fall into one another, erasing the safe distance that kept them in orbit. In this way, the resulting mass increases exponentially and begins to produce so-called "black holes." These are places with such a high concentration of mass that nothing can escape their gravitational pull. It is true, however, that this "collapse" is accompanied by the emission of energy no longer needed to maintain the distance between nuclei in the process of approaching one another. At the same time, in his variations on Michelangelo's themes, Blake thickens the pectoral muscles and the layers of fat, always treating them in a more schematic and generic fashion (as though he were a victim of an inadequate knowledge of anatomy and therefore compelled to adopt very simplified outlines). In effect, in his sketches the figures collapse toward the centre of the work, where they would lie inert were it not for the onset of a contrasting movement, i.e., the emission of bursts of energy otherwise unutilized. In fact, auras, darting tongues of energy, lines of energy, capable of delineating a field of radiation emanate from Blake's compact and massive bodies.

HEINRICH FÜSSLI, FRANCISCO GOYA, AND JACQUES-LOUIS DAVID

What has been said about Blake applies more generally to fellow countryman and almost contemporary, John Flaxman (1755–1826),[8] who exaggerates the two concurrent processes of flattening and abstraction. In his designs, bodies are emptied as though on a strict diet, and seem to cling to contour lines that flow with amazing sureness and fluidity, as though cut by scissors. No one more than Flaxman declared war on the dogma of modernity based on the cult of perspectival illusionism and the mysteries of depth of field. From this point of view, he appears as an anti-Leonardo in art history. If the great Tuscan came along to institute the technique of creating a sense of distance in representation, pushing the vanishing point into infinity, the Englishman accentuates nearness. The two-dimensional surface

8 For additional information on Flaxman and all the authors of the early 1800s discussed in the pages that follow, see, Robert Rosenblum and H.W. Janson's *19th-Century Art* (New York: Abrams, 1984); Hugh Honour's *Neoclassicism* (Harmondsworth, England: Penguin, 1968) and his book *Romanticism* (New York: Harper and Row, 1979). See also my *L'alba del contemporaneo* [The dawn of the contemporary] (Milan: Feltrinelli, 1996 and 2008). Some of the ideas developed here can also be found in the catalogues I provided for exhibits on the great illuminators of Dante, organized by C. Gizzi for the Casa di Dante degli Abruzzi (Torre dei Passeri, Pescara). See the catalogues *Füssli e Dante* [Füssli and Dante], *Blake* e *Dante* [Blake and Dante], *Flaxman e Dante* [Flaxman and Dante] (Milan: Mazzotta, 1985, 1983, 1986, respectively).

becomes an unavoidable end-point. We can say that, with Flaxman we have the practice of cloisonnism a century before this technique is definitively adopted by Émile Bernard, Paul Gauguin, and the entire group of Symbolists (the generation "born around 1860" who would reach maturity in the 1890s). The adoption of the rigorous standard of cloisonnism has its logical consequence, which is the adoption of an equally rigorous parataxis. Flaxman's empty shapes obviously cannot be superimposed upon one another because, if they were, they would no longer be distinguishable. This distinguishability is entrusted entirely to line, but the bodies have to be detached from one another, arranged side by side in a sequence, like a sarcophagus frieze, which is generally a feature of "primitive" epochs. These forms contrast with those of classical and mature periods, which tend toward dense compositions marked by hypotactic structure.

Having dealt with the strong binary of the precursors of contemporary or postmodern taste, we need to go back a few years to consider the members of an earlier generation who should be credited with being the first to make "the great refusal" with respect to modernity and with opening the phase of protest against that style. These figures are separated from one another by great geographic distance, professional status, and temperament; yet they can be brought together on the basis of the criterion of "long-distance collaboration" mentioned in the first chapter. In this case, the collaboration is made easier by the fact these "operators" belong to the same field, that of visual art.

The first of these innovators, first in terms of chronology and importance of personality, is the Swiss artist Johann Heinrich Füssli (1741–1825). In his work, a form of protest against traditional modern art (which still appeared toward the end of the 1700s in some aspects of the late Baroque, known as *barocchetto* in Italy, a style that in France is referred to as *rocaille* or Rococo) appeared in the very nature of his training. That training for a long time was unlike the usual art practices, whereby the artist came out of an apprenticeship in a workshop under an accomplished master and underwent a slow grooming in all aspects of the craft. In his native Zurich, the young Füssli acquired a rich culture in the form of religious studies (he was on his way to becoming a pastor in a Zwinglian church, a form of Calvinism), to which he added equally refined literary interests inspired by a great philologist of the times, Johann Jakob Bodmer, who introduced him to both classical literature (the Homeric poems) and German literature (the Nibelung poems and the Eddas) and exposed him to "Romantic" themes, which is to say, popular medieval themes. In all this, visual art initially played a marginal or supporting role for him, as was appropriate for a tireless traveler (visiting the great European capitals), always prepared to make sketches. This training, which was more suited to an intellectual person than to the average artist, permitted Füssli to break with the normal *consecutio temporum.* In other words, he interrupted his contact with the immediate past in the form of the late Baroque tradition in order to pursue his interest in earlier historical periods, which included the distant ancient Greco-Roman past and the more decidedly sophisticated Mannerism. A long "Italian" period (from 1770 to 1780) was decisive in the maturation of Füssli's artistic vocation, although for the entire decade he continued to produce designs at best enlivened by patches of water colour. In fact, he was able to see traces of the classical world and the noble, passionate Renaissance in Michelangelo, later developed by the Tuscan Mannerists; but Füssli's visual education during his Italian period was truly extensive, so much so that it almost constituted his entire private handbook of the history of art, which included contact with the picturesque (the colour of the Venetians, the naturalism of Correggio in Parma and the Carracci family in Bologna). For the time being, however, he preferred the distorted, artificial language of the Florentine Mannerists, which consisted of elongated, surging figures. Even the Baroque tradition indulged in exaggerated, dynamic figures in agitated scenes, but always within the framework of a plausible observation of natural atmospheric phenomena approached on their own terms with a genuine fidelity to optical techniques. Instead, the upsurges that Füssli gives to his mannequins in his "Italian" designs indicate an excess of energy that we would later find in Blake. The source of this feature is mysterious and ill-defined if we do not presuppose the existence of a connection with the newest resources of electromagnetism. These are "different" energies than those of the well-known verisimilar-naturalistic kind. As will be the case with Blake, these energies tend to be concentrated on the human subject rather than dispersed into the cosmos. Thus, they assume a psychological quality that reveals disturbing upward surges in the human subject. In short, we have the idea that our psychic life is split, that it unfolds at two levels. One corresponds to the waking state and has a well-established function. Essentially, the philosophical schools of Rationalism and Empiricism, which converge in the Enlightenment, had attempted to explain the mechanisms and functioning principles of that state, but in the very act of dedicating great analytical attention to it, they also proceeded to "remove" (Freud would later say), repress, or deny the dark continent of sleep and dream. Similarly, Galilean-Newtonian physics had limited itself to analyzing only data of the physical-material realm.

The time has come to deal with those dark spaces where the physical and metaphysical principles that elude the established order ferment, where figures leap, arch their back, recede from the foreground to the background, disregarding the laws of perspective. We no longer have the mathematical proportionality of the relationship between bodies, as laboriously constructed by the modern tradition. Long, dramatically foregrounded bodies are flanked by dwarfed figures that recede into distant horizons. This is similar to ether, which contains electrical phenomena (the so-called "dialectric" in the terminology of that period), has no consistency, is not measurable, and is an entity that joins rather than divides or segregates, obeying the laws of attraction and inclusion rather than separation. The figures in Füssli's drawings and paintings seem ready to be linked together as in a chain or like the cells of the voltaic pile, which are aligned so as to create a difference in potential among them, which permits a current to flow within that circuit. This is also similar to the chains formed by holding hands in séances, said to help the "fluid" or spiritual force circulate, even if we are moving from the field of physics to that of the preternatural, in this case.

After his "Italian" period and a brief return to his native Zurich, Füssli takes up residence in London where he lives until his death (he also anglicizes his surname to Fuseli). He challenges the domination of Joshua Reynolds as well as the ideals of the Royal Academy, into which he is however accepted because his position appears to be less provocative and scandalous than that of the younger Blake, with whom he nevertheless dialogues on the basis of a shared sensitivity toward the "other" or the "different."

A few years after Füssli, Francisco José de Goya y Lucientes (1746–1828) is born in a small Spanish town near Zaragoza. On the surface, the two careers seem to be incompatible. Goya is a true working man, a craftsman with a solid artisanal tradition. Though he manages to escape provincial life, settling permanently in Madrid, he does so in order to engage with a modestly utilitarian and applied activity. In fact, he is commissioned by the Bourbon royal family to provide sketches and cartoons to be transferred to tapestries, intended to brighten the private apartments of the king and the Infante. But there is an ironic twist in all this since the practical function of the tapestries helps Goya to design his images in two-dimensional, flat style, which, as we have seen, is one of the dominant features of the impending contemporary age. The tapestries are coarse-grained handicrafts and call for an art consisting of clear shapes and contour lines as well as patterns that are almost cloisons. There is a structural affinity between these and the many other forms of popular images that already existed in other eras (Gothic stained-glass windows, mosaics, and glazes) which in turn can represent a bridge with painting techniques popular in the contemporary or postmodern era: the affiche, the comic strip, the cartoon, and the polychrome forms of video games. In other words, Goya finds himself almost unwittingly compelled by technical reasons to develop those surface qualities (of a closed, linear, clear, paratactic art) that would return with full force in one of the early stages of contemporaneity with Symbolism and its dual registers, represented respectively by Gauguin (using large, synthetic units, like the panels of a Gothic stained-glass window) and Seurat (using many small units, like the tesserae of a mosaic or the fibres of a tapestry). In both cases, the primary fact is that the image is born in the workshop and is made of a limited number of discrete components arranged in a calculated, rational fashion. In a sense, we are dealing with a low-definition image, which is quite different from the modern-naturalistic one of photography; therefore, the image tends to be located at the abstract, generalizing level because of its nature. These are the characteristics that Blake arrives at by taking a different route.

The foregoing pertains to the constitutive aspect of the images. As regards their psychological connotations, Goya adopts the same double register we find in the paintings of Füssli's London period: flat, stylized scenes meant to be transferred to tapestries. These are festive, even cloying, depictions of the pastimes of a population of *majos* and *majas*, young commoners living alongside the aristocratic and middle-class world to which they bestow the gift of their youthful innocence; except that the "other" half of our psychic life takes control. A "black" register emerges from the interstices of the scenes, slowly replacing each sunny element with its nocturnal counterpart. In other words, Goya too proudly depicts the Freudian "return of the repressed." The sleep of reason evokes monsters, but it is good that this happens since it allows the submerged continent finally to emerge. In doing so, it can now avail itself of the guidelines prepared by the world of the waking state. Those flat images arranged in a series, which invariably remind us of comic strips, are extraordinarily useful for the representation of long narratives of nightmares. In his final phase, when around 1820 he paints the walls of his modest dwelling outside of Madrid, called by the people "the Deaf Man's Villa," Goya confirms a decades-long coherence with his early compositions, which are arranged horizontally, almost like a frieze, in a paratactic sequence similar to enormous wall scrolls. Now, however, he places a series of delirious, monstrous beings in those strips, which are not very different from the forms being produced in the same period by his homologue, Füssli, even if there was no direct contact between the two.

The third "great" figure among this group of proud initiators is also the most conformist, the most in-tune with the reception possibilities of his time: the Frenchman Jacques-Louis David (1748–1825). In his case, we even have an official label, Neoclassicism (whereas there are no appropriate labels to designate with a single term the innovations of Füssli and Goya). But this is a familiar label and its currency is well established, a fact that apparently keeps it removed from the strange, unusual homologies that I, instead, am trying to make with the history of late eighteenth-century style. In a sense, Neoclassicism pushes the limits of the preceding regime centred on late Baroque and Rococo; but its impact is contained. It is one of those recurring "call to order" movements that surface in the history of western art whenever there is a tendency toward exaggeration, as occurred in the *barocchetto*, with its swollen, exaggerated forms on the verge of coming apart in a haggard picturesqueness. Essentially, Wölfflin's "binaries" are sufficient to explain that reversal of taste, which runs to adopt closed, linear, superficial forms against the excesses attributable to artists like Giovanni Battista Tiepolo, Jean-Honoré Fragonard, Joshua Reynolds, and Thomas Gainsborough. Up to this moment, swings of the pendulum appear to be contained within the field of art. What do the epochal changes we also find in material culture have to do with all this? And how are these movements connected to the electromagnetic revolution? David is not merely a reincarnation of Raphael or Poussin. If he were, we would have a "return" that is perfectly classifiable within a cyclicality that is endogenous to artistic matters. At least in his heroic phase, which corresponds to the last two decades of the century, when he produces his painting manifesto of Neoclassicism, *The Oath of the Horatii*, David goes beyond a simple revival of harmony and grace that can be referred to generally as neo-renaissance and classical. In his work, we find a bold experimentation with the flattening of space or reduced depth, which links him surprisingly to his brethren of lesser fame, Blake and Goya, as well as Füssli to a certain extent. David's composition is boldly schematic and reductive to the point where it would perhaps not have met Vasari's criteria for his third *maniera*. In other words, even the David of the "appropriate" years enters the idealist club of the Pre-Raphaelites, those who reject the chronological start of modernity represented by the Vatican rooms and rediscover the archaizing schemas cherished by earlier generations, the protagonists of a fifteenth-century fond of pure and linear solutions, the same which art historians north of the Alps tend to characterize as "primitivism."

This excursion through the great precursors of the contemporary or postmodern period would be incomplete without some attention paid to many others who belong to the generation of Blake and Flaxman. We should remember, for example, the Danish artist Asmus Jacob Carstens (1754–1798) and the Italian champions of the new styles, who were also very important in the order of their respective dates of birth, including Andrea Appiani (1754–1817), who is close to David in terms of the practice of a reductive, primitive Classicism inspired by the paratactic arrangement of the stories narrated on the columns of Roman antiquity (Trajan's column and the column of Antoninus Pius), which in turn are almost forerunners of the "strips," sequences, and arrangements of contemporary narrative genres, such as the comic (the most appropriate term to describe this genre, is indeed strip). The need to frame the narrative in narrow horizontal bands forces the artist to reduce the profiles of soldiers (for example, Napoleon's troops, whom he extols in the frieze for the room of the Caryatids in the royal palace in Milan, unfortunately destroyed during the Second World War). In this, Appiani spontaneously rediscovers the styles dear to Goya as well, namely, the cartoons for tapestries.

Still following the chronological order, I need to mention Antonio Canova (1757–1852), the imitator and counterpart of David in representing Neoclassicism at the highest level as sculptor. But in addition to this very famous sculpture, the Venetian artist produced a number of studies, sketches, and tempera paintings that seem to be very close to the distorting, abstract, primitive tendencies we encountered with the other members of the same generation, especially Füssli and his contemporary, Blake.

Finally, we also have to mention Felice Giani (1759–1823) who illustrated very well the neo-mannerist taste in series of rich wall frescoes, displaying his figures in unnatural, artificial poses of great sophistication as regards forms coloured in the same strident chromatic schemes we saw in the work of Füssli and Blake. But it is not my intention here to write a detailed history of these intense and revolutionary decades. This task has been carried out in a study where I dealt systematically with the "first" contemporary age, which is the distant initial phase of postmodernity (let us say the period 1770–1860) in order to bridge the gap that extends up to the emergence of the clearly innovative work of Paul Cézanne within the decade 1860–1870. The period that stretches from that point to the present day has already been examined with a certain degree of analytical rigour in my work *L'arte contemporanea* [Contemporary art], but it needs to be qualified by a limiting modifier, such as "recent" or "second phase," to distinguish it from the first, foundational phase. With respect to this 200-year time span, I limit myself here to discussing only some key moments or thresholds crossed in an attempt to bring these stages into a general culturological framework.

A DIALECTICAL PAIR: J.M.W. TURNER AND DAVID CASPAR FRIEDRICH

In the context of these particularly noteworthy and substantial developments, the present study could not do without mentioning Joseph Mallord William Turner (1775–1851), born approximately one generation after the protagonists discussed above. To begin with, his work is characterized by a clear shift in genre relative to that practised by other artists who, as we have seen, dealt primarily with the subject of the figure and history. The innovative features I have pointed to in their work reside in the way the artists treat the human figure and insert it within the syntax of the narrative: reduced, stylized, flat images tending toward abstraction to the point where they are reminiscent of the style of the "primitives," thereby eliding the entire cycle of modernity and its principle of fidelity in representation. To schematic abstraction they add bursts of energy and contortions derived from the anti-modern style that was Mannerism.

Turner,[9] instead, is almost exclusively a landscapist, focusing on expanses of land, sea, and sky. This enables him to produce clear, precise wave forms, which are not difficult to associate with the waves of electromagnetism and all the other curvilinear designs dear to the new discipline of electrology. Turner's *vedute* or views exclude straight lines and lines that meet at right angles. Even the principle of single-point perspective, the pillar of the "modern" vision, is attacked. The English landscapist undoubtedly places "foci" in his paintings. They are points of great luminosity, like discs that may represent the sun and the moon but which are not limited to a single unity. Even though exploiting something like mirror images, reflections, and duplicates, the artist always places two or three foci (in the optical-geometric sense of the term) in his composition. Thus, we abandon the "centric" curve which is traced around one unique focus (the circle, which is too regular and static) and enter the realm of "eccentric" curves (spirals, hyperbolas, etc.) which need to be constructed through many foci. This is exactly like the waves used in electrology when we try to give visual form to phenomena that are in themselves invisible, at least in terms of the limited threshold of our sensory apparatus. In other words, Turner, perhaps unconsciously, applies the concept of "the centre is everywhere." In a contemporary universe governed by the very high speed of light, it is pointless to want to fix a source of light since it can fade or undergo change. We may recall that another basic property of our culture is the mutual convertibility of matter and energy with two consequences, both of which can be found in Turner's paintings. Firstly, the origin of light is now endogenous, that is to say it comes from within matter itself. It is naïve to assume that it emanates only from the sun or a secondary source like the moon. Secondly, following from the preceding premise, every point of the universe, every zone in the panoramic view can be the site of an unexpected burst of energy. This is why Turner always places more than one focal point in his landscapes and, more importantly, places them with great freedom, much to the viewer's surprise. It goes without saying that the light emanating from those unexpected

9 Among the numerous studies on Turner, it is worthwhile mentioning that of Francesco Arcangeli in *Dal Romanticismo all'Informale* [From Romanticism to the Informal] (Turin: Einaudi, 1977), in addition to the relevant chapter in my *L'alba del contemporaneo* [The dawn of the contemporary age] (Milan: Feltrinelli, 1996 and 2008).

sources is always more powerful than what seems plausible when the sun is taken as the point of reference. The glows in Turner's artwork are exceptional, like those of a fire or volcanic eruption, but perhaps we should equate them to the sinister light of a nuclear explosion in order to understand them and to provide a reasonable equivalent.

From the same generation as Turner is the Danish-German painter David Caspar Friedrich (1774–1840), who somehow attains a perfect union of these two subjects, the figure and the landscape, applying the best innovations I identified in the works of his predecessors and contemporaries. His figures are stylized and inscribed in a perfect, archaizing manner where flowing lines uncontaminated by atmospheric defilement triumph, and even the lines of the landscape have a fluidity that can be found in the paintings of Turner, as evidenced by the predominant use of undulations. In a word, even Friedrich can be accorded membership to the idealist club of the Pre-Raphaelites, that is, those who are prepared to abandon the sensuality, the moods, of "modern" painting in order to return to a pure, uncontaminated fifteenth century. But perhaps Friedrich puts too much coldness and rigour into his canvasses; after all, we are getting close to the point where we have to ask ourselves if it is enough to go back to the 1400s and stop there in order to find evidence of the onset of contemporaneity. All the precursors about whom I have spoken tend to go further back in history, to the "primitive," in order to find abstract processes, unrestrained schemas capable of transcending the "modern" love of rich, detailed representation. In other words, the "finished" product can prove to be a sort of *caput mortuum* [worthless remains], especially if practised with the attention to detail of a Flemish artist or a member of one of the many generations we find in the Italian fifteenth century, while remaining within the boundaries of Vasari's second *maniera* (for example, Andrea Mantegna, Sandro Botticelli, and Luca Signorelli). We must also recognize that the precursors mentioned above not only kept their distance from the pedantic, "finished" product, but they challenged it with energy and an abundance of "unfinished" touches, perhaps leaping forward a few years to borrow the spiral, serpentine forms of the Mannerists.

Friedrich is different from them on both fronts in that his work is overly "finished" or detailed, and thus he shuts himself off from the infinite, the *élan vital* as Bergson would say. It is not happenstance that one of his favourite subjects is the winter season: a vision of frozen, wintry expanses or masses of polar ice that capture ships like a vice and crush them. In a word, the pure whiteness of Friedrich's paintings contrasts with Turner's tumultuous seascapes or with the tense, exaggerated poses distorted by impetuosity that we find in the figures of Füssli, Blake, and David himself. On the other side of the coin, it is also true that Friedrich bypasses the Raphael effect; so how can we deny him entry into the ranks of the post-moderns who are intent on creating a bridge of affinity with the pre-moderns? If the dividing line is to be the frescoes in the Vatican rooms, surely Friedrich ignores their lesson in freedom and fluidity and pretends either that it never existed or that it could be erased from the history of the West.

BACK TO THE MODERN

The preceding remarks are even more relevant to some members of a later generation. Moving toward the end of the century, as regards the date of birth, we find the figures associated with the German Nazarene movement; for example, Franz Pforr (1788–1812) and Friedrich Overbeck (1789–1869). The case made above for Friedrich applies to them as well. If all we need to do is go back to the 1400s to find the start of the contemporary age, leaving out Raphael's work in Rome and most of the Classicism of the first part of the 1500s (Vasari's third *maniera*), then all the members of the Nazarene art society must be included. In part this is because they reject impure atmospheric effects and tangled compositions that simulate the horrible randomness of existence or, more generally, the lowered standard that accepts our earthly drama, such as it is, without making any effort to ennoble it. The "verisimilar" Raphael is also the symbol of a desacralized western ethos that no longer believes in transcendence, but accepts the primacy of nature and sets out to know and dominate it, as well as to intervene in it without mental reservations or half-measures. Against all this, the Nazarenes take the road of "protest" that affects art and visual style; above all, they mobilize the general principles of custom and morality. For them, to dress in the Nazarene style means to rediscover the values expressed in the figure of God-the-Son, Christ, with his purity, and to reject possessions, the temptations of commerce, and complete immersion in the worldly trappings. This is a luminous road that would be taken by many avant-gardes throughout the contemporary age. A few decades later, it would be taken by Dante Gabriel Rossetti (born in London in 1848) and his brotherhood, who would finally coin the very effective label of Pre-Raphaelitism, pronouncing a word that was already in the air; in fact, it had been announced and for the most part practised by many earlier generations. (All the innovators discussed up to this point were Pre-Raphaelites *ante litteram*.) Then, we have the revival of these same qualities by Paul Gauguin and the other members of the Pont-Aven School, and continued by the Nabis, one of the groups forming the array of Symbolist movements. Taking another leap forward (without mentioning the historical avant-gardes of the early 1900s), we arrive at the second half of the 1900s where we witness an inexhaustible series of "reverberations" of the anti-west protest: the Beat Generation, the 1968 movement, counterculture, the 1977 movement, and so on.

We must also recognize another contribution by Overbeck and the other Nazarenes. They were opposed to the ambiguity of the preceding "official" avant-garde, i.e., Neoclassicism, about which I have already spoken. How are we to understand the project of David, Canova, and their followers? The project was truly innovative if it sought "primitive" values in remote Classicism, in other words, if it found the courage necessary for abstract synthesis or if it uncovered the source of spontaneous energy. It was not innovative, however, if it was obsessed with the cult of the canonical models of Greco-Roman Classicism in versions that are a bit too antiquated, or by the similarly "high" culture revival of Raphael, understood to be the founder of an illustrious "series" that was subsequently praised by Annibale Carracci and Nicolas Poussin. Seen in this light, Neoclassicism (as already noted) was nothing more than a swing of the pendulum occurring completely within the sphere of art: a corrective, counterbalancing motion, a break with the excessive "openness" of the late Baroque and Rococo. In ethical, ideological terms, all this may have seemed to be a return to the austere models of Roman-ness, but these models are mummified in the sense that they are transformed into a suffocating blanket that could be easily appropriated by despotic regimes or tyrants – from the relatively "illuminated despotism" of Napoleon in his "excursion" through the French Revolution, to the much less dynamic authoritarianism of the regimes restored following the Congress of Vienna. These figures, however, found it useful to display their general spirit of conservatism by resorting, in aesthetics as well, to the rigid canons of proven effectiveness, such as those of Neoclassicism. This "ism" was like a historical missile: the first stage had a beneficial effect removing the wreckage of the old regime, while the second stage, on the contrary, acted like a brake, or like a "cork" that prevented new energies from bubbling forth. It is precisely from these *aporias* in the neoclassical style that, by contrast, a style emerges corresponding to Romanticism, understood in one sense. This is certainly not the only possible meaning, nor is it the most positive or most advanced, according to the parameters set down in the present study.

Considering its etymology, the term Romanticism signals a change in theme and content with respect to sources from which to draw inspiration. These are no longer conventional, canonical, ossified, Greco-Roman Classicism and its Renaissance version, but rather the material that this Classicism had contributed to "removing:" the intervening centuries, the Middle Ages, with their splendours, myths, and stories narrated in the Romance and Germanic languages. The period is a repository of fables, in the broad sense of the word, that are definitely fresher and more exciting than stuffy Greco-Roman epics.

But there was the danger that this decidedly more appealing material might justify a return of the modern cast of mind, consisting of attachment to facts, to accurate and detailed descriptions, whereas Neoclassical storytelling, precisely because of its *terribilità* [frightening power] and its high degree of conventionalism, turned away from these intimate prose narratives. In a way, a risk of this sort can already be found in the art of the Nazarenes, which certainly falls within the archipelago of Romanticism, at least at the level of theme, in as much as it seeks to tell medieval stories or stories related to the Christian *epos* and its *ethos*. Undoubtedly, the epic that the Nazarenes offer up in their paintings is fresh and light, and it points to the possibility of a great, new beginning in a purer, more innocent world. Above all, an aesthetic element is added to the contents of the painting, and this is precisely the moderately "primitive" tendency to look back to art "before Raphael." In the meantime, however, the narrative becomes detailed and dense. Theirs is a primitivism that functions like a Trojan horse or like germs that cause the resurgence of the "illness" of modern illusionism. We know that an antibiotic regimen is ineffective unless it is fully implemented to destroy the harmful bacteria. In a body that regains its strength, surviving germs can multiply with more virulence than before and regain their hold in a short time. Similarly, the "finished" painting executed by the Nazarenes with obsessive precision acts like a resurging infection, unleashing a second phase of mimetic, representational modernity set to develop in forced stages but with great intensity and vividness, as far as the splendours of Realism and Naturalism in the 1900s.

Moreover, the birth of the leader of the Nazarenes, Overbeck, comes a mere nine years after that of Jean-Auguste Dominique Ingres (1780–1867), who is David's best and most famous pupil. It would seem, however, that he served on another front, that of austere, inflexible Neoclassicism, which is fundamentally pagan in its customs and leans toward a granitic cult of the beautiful, from a strictly thematic point of view. No one can compare Ingres' noble, elevated, and at the same time profane subjects with the pure, virginal, Christian vignettes executed by the Nazarenes. But from the standpoint of style, we can say that the artists exhibit a common attention to vivid detail. With Ingres, we are now quite a distance from the primitivizing abstraction displayed in the early canvases of his teacher David, although the subject is fanatically classical (Greek or Roman); but this is the beginning of a process of reducing and flattening the image, and of inscribing the story on a flat surface, as discussed earlier. It is true that the last works of David, at the time self-exiled in Brussels, hints at a gradual reintroduction of details, producing designs that are increasingly solid as he strives to restore the roundness of objects and bodies. The pupil Ingres certainly goes further in this direction as he reduces the scene. His compositions are not large swaths of episodes narrated in a free, paratactic language, but rather sharp close-ups of full, real elements; flesh and blood figures, items of clothing, jewellery, decorations, and accessories, all receive incisive treatment and are rendered in perfect focus. We could say that we are still within the revival of a mindset that can be described as "before Raphael" but which, by the end of the 1400s, appears in the work of Botticelli and Signorelli, as well as their followers, who were thirsty for reality, indeed, hyper-reality, but were unable to give bodies the magical touch that could set them free and allow them to move "gracefully;" in other words, those artists were Pre-Raphaelites or "primitives" almost reluctantly, because the time was not right, rather than by a conscious choice of the roles they wished to play. They left behind purity, abstraction, and the essence of forms, which had been the features of someone like Piero della Francesca, Masaccio, or Paolo Uccello (and going further back in time, artists like Giotto and Cimabue). In other words, we have to distinguish between two kinds of primitivism. The one we find in the analytical and well-defined works of the Nazarenes and Ingres is primitivism at the minimum level, which is found only in the fact that the corrosive acid of the atmosphere does not envelop a well-defined representation of this sort. Bodies and objects are positioned as though under a bell-jar or in a vacuum, so the gaze can scan them without filters, as though it were adhering to the skin and, in a sense, overlaying on it a coat of insulating, protective varnish that makes the fabrics glitter, as is done to preserve archeological finds entrusted to museums for eternity.

But why stop at that difficult, untenable, middle position – one foot in and one foot out – in relation to the canons of "modernity?" A step beyond is taken by the members of a slightly later generation to which we must recognize the leading role played by some French artists destined to acquire great prestige and to be regarded as the founding fathers of "modernity," a term used to designate what, for the most part, I have insisted on calling the contemporary age. Théodore Géricault (1791–1824) and Eugène Delacroix (1798–1863) see as a pointless and painful, self-imposed limitation the goal of Ingres and the Nazarenes to stop at the threshold of Raphael's maturity. What escapes them completely, however, is the fact that these limits are consciously accepted by slightly older artists as the grand strategy of opposing modern, verisimilar representation and of reconnecting with the sources of pre-modern archaism. In a sense, our two artists relive the situation in which Leonardo and his heirs (Raphael, Giorgione, Titian, and Correggio) found themselves a few centuries earlier; the main issue was once again that of looking past the hardness of the "second *maniera*" in favour of the softness of the third. They re-experience, at an accelerated rate, the stages that led Raphael himself to surpass the vividness, staticity, and symmetry of the compositions he produced while a pupil of Perugino to develop the more picturesque, atmospheric, spontaneous style we find in the Vatican rooms. But it was necessary to go well beyond in this second phase, which involved an accelerated revival of a whole segment of historical evolution. In fact, the same divine Raphael had not been able to avoid being stuck in some residual, fifteenth-century hard-edged painting. It would take the next phase of the 1600s, the Baroque age, to complete this transition. Only a century after the achievements of Raphael, the splendid modern artists would emerge who could perfect verisimilar, "natural" representation: Annibale Carracci, Caravaggio, Peter Paul Rubens, Giovanni Lanfranco, Guido Reni, Pietro da Cortona, and Nicolas Poussin, who shared a general objective despite the many differences in personality, temperament, and school. Géricault re-proposes the sanguine, plastic, atmospheric styles of a seventeenth-century *koine* that best fuses the achievements of the leading figures mentioned above. But in doing so, he is not aware that he is reversing the course pursued by David in his day. This is a David who started out precisely with the "conventional" Classicism (which is to say, natural and noble at the same time) of the Carracci family or Poussin, but abandoned that style in his quest for poor, leaner, more intellectual images that looked to the past. Once this retrospective tendency comes to an end, we see the resumption of a progress that is almost "embarrassed" by the stoppages it had to endure and moves quickly to cull the best contributions of the seventeenth century to realism, and finally be free to produce a totally "modern" chapter of art history, one that is in direct contact with the most burning issues of the day.

What illustrates this point magnificently is a news item, a *fait divers* the French would say, like the sinking of the ship Medusa and the accompanying scenes of panic, horror, and perhaps cannibalism among survivors clinging to the famous raft. This is Théodore Géricault's first major work, a grand painting of "modern" realism precisely because it concedes nothing to the stirrings of abstract and stylized postmodernism, which appeared in the work of his predecessors in France and elsewhere. By contrast, in this painting, the bodies are full, solid, and stout; they suggest a masterful knowledge of anatomy and interpret the moment in spectacular fashion, as if surprised by the flash of a camera, or as though constructing the set of a *tableau vivant*, satisfied in their ability to render the atrocity of an action caught at its most critical point. Stated differently, *The Raft of the Medusa* (1818–19) is rich in motives and aspects that contemporary art would want to destroy, dismiss, or refute (and had already tried to do so through the precursors mentioned above).

Eugène Delacroix moves in the same direction by intensifying, if that is possible, the agitation of the scene, the tumult of the bodies, and the atmospheric effects. If the language of Géricault is grandiose Baroque naturalism, the language of Delacroix revives aspects of the late Baroque and Rococo, but in this respect the artist moves further away from "primitive taste," leaving no trace of it. The "unfinished" work wins out, but it does not possess the inspiration that blows through the paintings of Turner. What we have, instead, is the complete triumph of the picturesque, the illusion of depth, and stormy atmospheres in an agitated composition; in short, the definitive reappearance of the "open" form of Wölfflin's binary, to use once again the frame of reference adopted in the present study.

It is worth recalling that this second cycle of triumphant naturalism finds an important enthusiast in Italy as well, in the person of Francesco Hayez whose date of birth, 1791, is not far removed from that of the two great Frenchmen. (He lived an extraordinarily long life, dying in 1888). In his case, too, we have the curious process of "doing and undoing," the sudden reversal of the hourglass, which we encountered in Géricault and Delacroix. Consider the fact that one of the reasons for the anti-modern revolt of David, Blake, Flaxman, etc., had been the desire to reassert the primacy of "Tuscan" draftsmanship over "Venetian" colour. But we know that it was not an empty claim to return to an ideal, atemporal, eternal Classicism, in which forms would be nobler if referred to the sensuality of the flesh. For these artists, the primacy of the design was basically the assumption of an abstracting, simplifying attitude, as is appropriate if the examination of objects is made by using waves and not optical rays.

Hayez either misunderstands or does not understand this as he engages in simplistic reasoning. If there is be a "return," why limit oneself to using the essentially poor and not very rewarding resources of Tuscan Classicism (Raphael and Michelangelo) and not the more appealing, attractive, mimetic resources of the Venetian school, from Titian to Paolo Veronese and Giambattista Tiepolo? In the case of Hayez (as with Géricault and Delacroix), history is not used to check the tendency toward "verisimilitude," the mimetic realism of modernity, and to rediscover its earlier phases, but rather to remove the barrier of Neoclassicism in order to restart the movement toward the ideal of a representation that is always driving, active, and sensitive, even if this same "historicism" would always deny Hayez, and every other member of his generation and generations immediately following, the possibility of realizing a representation of the real in its quotidian guise. What remains is the delaying and diverting mediation of costume scenes, but these costumes are medieval in flavour and, therefore, belong to the *ethos* of Romanticism, according to the surest and most verifiable meaning of this term, which is a thematic one. If we turn to Italian literature of the same period, we find a writer who, much more authoritatively than Hayez, describes the same scenario of a modernity that is arrived at through the "Romantic" illusion, the re-enactment of history. I am obviously referring to Alessandro Manzoni, almost a contemporary of the Venetian painter (born in 1785). His masterpiece, *I promessi sposi* [The betrothed], is set in a specific historical period, i.e., the time of Spanish rule over the city of Milan, and it describes morality and customs that can be referred to as seventeenth-century Baroque. But it does so for the purpose of representing themes, mass movements, and problems of individual and collective psychology that could not be more "modern" in every sense of the word because they are tied to the present that Manzoni himself lived. This is the Milan of the first part of the 1800s brought to light with objectivity, analytical care, as well as the lucid determination to understand the great socio-economic, ethical, and political processes – a project that corresponds to the aims of the great "modern" narratives. From this point of view, Manzoni should be compared with the French writers who created the "modern" novel, which is dedicated to describing the rise of the bourgeoisie and the establishment of a set of values decidedly laic, scientific, and experimental in nature; I mean writers like Stendhal (1783–1842) and Honoré de Balzac (1799–1837).[10]

The Romanticism we could attribute to the other great figure of Italian literature, Giacomo Leopardi (1798–1837), is very different since, in his case, the thematic, historicist data should disappear entirely. Here we have a refusal to use medieval subjects as "Trojan horses" i.e., the analytical principles of representation. What counts for Leopardi is the infinite "breath" of a primordial energy, which is very much like that intuited by William Blake: an energy that coincides with the infancy and adolescence of humanity, when humans are free to express their imagination spontaneously and when the restraints of reason, society, and community are not yet in place. The result is that we find a sincere, inexhaustible cult of primitive ages and origins in Leopardi. He understands that the medievalism so dear to committed Romantics with their lugubrious, nocturnal themes, are pretexts used to promote a literature that suffocates due to its excessive reliance on description and attention to detail. It is very difficult to find in such a narrow timespan, two great writers more different than Manzoni and Leopardi. The first was an effective "modern" (a promoter of the illusion of reality in a clear-eyed analysis of the society of his time); the other was an equally strong supporter of all the

10 On Manzoni, see my *Dal Bocaccio al Verga. La narrativa italiana in età moderna* [From Boccaccio to Verga: Italian narrative in the modern era] (Milan: Bompiani, 2005); on Stendhal and Balzac, see another essay of mine titled *La narrativa europea in età moderna. Da Defoe a Tolstoj* [European narrative in the modern age: From Defoe to Tolstoy] (Milan: Bompiani, 2010).

possible arguments for being an anti-modern, someone who dreamed of a return to a period "before modernity," which can be converted into possible signs of postmodernity.[11]

Naturally, Leopardi knew full well that he was a loser in this regard, at least in the years immediately following, which would be dominated by the social prejudices he liked to define as "magnifiche sorti e progressive" [the magnificent progressive destinies]. We would have to wait for the return of the contemporary with full force at the end of the century for his ideas to re-appear. In short, three generational waves confirm this return of modernity after the first violent but precarious prelude to postmodernity, of which Leopardi himself was a promoter. We have to refer to those born around 1800, followed by those born around 1820, and finally those born around 1840. Only with those born around 1860 does the postmodern syndrome start up again in the guise of Symbolism (and not without taking with it many of the residual effects of nineteenth-century Naturalism; but about this I will have more to say in the coming pages).

In order for the generational concept to be applicable in all these cases, we have to accept the fluctuation of ten or more years mentioned in the previous paragraphs, so even a Géricault, a Hayez or a Delacroix can be considered to belong to the wave of artists born around the year 1800 (fitting in almost perfectly is a champion of literary modernity like Balzac, born in 1799). I would like to point out that, if we accept these flexible parameters, even Jean-Baptiste Corot (1796–1875) falls within the statistical average. In effect, his role was that of preparing a bridge from seventeenth-century landscape art (that of Poussin and Lorrain) to the pure landscape painting of the Impressionists, which is to say a golden bridge from one phase of modernity to another.

FROM ROMANTICISM TO REALISM

As a general observation, I would like to underscore the fact that, during this period, we witness a reversal in the function of "history" as

11 For a more detailed comparison of a modern Manzoni and a postmodern Leopardi, see my article, "Il problema della lingua nella narrativa italiana moderna," in *L'italiano in America latina* [The Italian language in Latin America] (Florence: Le Monnier, 1987). See also *Dal Boccaccio al Verga* [From Boccaccio to Verga] (Milan: Bompiani, 2003).

compared to what happened in the case of the pioneers of postmodernism. I made this point in my earlier comments on Leopardi, for whom nostalgia for origins and the past serves to announce (or prefigure) pre-postmodern qualities, a foretaste of the contemporary age (rejection of detail and naturalism, formal simplification bordering on distortion). Here by contrast, due to their freshness, chronicle-like quality, and ability to capture emotions and daily life, historical themes – mostly medieval – function like Trojan horses (as stated earlier) to produce the illusion of reality, thereby overcoming academic reservations in this regard. However, the members of the generation I am describing, from Hayez to the "young lions" of the Macchiaioli movement, such as Telemaco Signorini, struggled to free themselves from these same "historical" motifs shortly after adopting them. These historical motifs worked almost like catalysts in chemical reactions, which, as is well known, only accelerate or make easier a reaction without becoming an integral part of the chemical structure, so that, once the desired synthesis is achieved, the catalysts can be removed. The main objective of the art of the first part of the 1800s is precisely to free itself of any residual attachment to historicism, even though that attachment was useful and desirable initially. Nonetheless, in all the works produced in the intervening years, historical themes and period costumes, vignettes, and casts of mind, invariably play an important role. Only at the end of this span of years, with the young artists born around 1840, do we have pure, natural spectacle purified of all hesitations, constraints, narrative agitations, and edifying aspirations. But we have to be careful about the dates and not expect artists born around 1820 to have a goal that would be attainable only by their heirs born two decades later.

More specifically, this means it is appropriate to expect a fresh, true-to-life naturalism that captures reality using the technique of the "open window" from someone like Claude Monet, the leader of the Impressionists, born in 1840. The same can be said of his associates of the same age, such as Alfred Sisley and Auguste Renoir. It is more difficult to expect this result from artists who are slightly older, such as Camille Pissarro (1832), Edouard Manet (1833), and Edgar Dégas (1834), whose work is unavoidably a compromise with the vignette or subject, even though these artists produce original and creative compositions. We can find the same degree of "vignette contamination" in the work of artists in other countries without needing to say that there is a lag with respect to French art, considered to be more advanced than its European counterparts. In Italy, as noted above, Telemaco Signorini (born in 1835) easily manages to cleanse his landscape art of vignettes and especially period costumes. He succeeds in capturing the times in which he lives, representing them directly without much difficulty. This cannot be said of his older colleagues from the same group of Macchiaioli, Giovanni Fattori, Silvestro Lega, and Vincenzo Cabianca who, because they were born in the 1820s, are forced to deal with history as an obligatory stage on the way to representing that which is "true." The same must be said for Neapolitan artist Domenico Morelli, who can be thought of as Hayez's heir, continuing where his "elder" (older by almost thirty years) stopped. He emphasizes historical reconstructions (see his paintings, *The Iconoclasts*, *Torquato Tasso*, *Eleonora Duse*, etc.), and depicts the reality unfolding before him in all its dynamism.

Furthermore, this obligatory reflection on the history of painting is not a deplorable "lag" to be found only in (and considered a handicap to) the more advanced French developments. We can find this lag in the countries beyond the Alps as well, if we look further afield than the well-known cases of Jean-François Millet (1814–1875) and Gustave Courbet (1819–1877), whose work (especially Millet's) in any case is also very distant from the technique of the "open window" since these artists were capable of a controlled rhetoric of the emotions, states of mind, and poses; but certainly their repertoire does not include historical themes. In the French context, we also have other important artists, including Rosa Bonheur (1822–1899), Pierre Puvis de Chavannes (1824–1898), William-Adolphe Bouguereau (1825–1905), and Jules Breton (1827–1906). In each case, we are dealing with a hybridization or an amalgamation of slices of reality, vivid realism, almost *trompe l'oeil*, as well as historical recreations and period costumes, executed to satisfy the desire for detailed and sensual stories.

We should include in this category the great protagonists of other areas of Europe, for example, the Swiss-German Arnold Böcklin (1827–1901), whose eclecticism corresponds very well to a kind of superimposition of different images or interpretive keys. He delves into the mythic past narrating stories of fauns and nymphs, but this does not mean that he attenuates or simplifies the representation. Instead, he fills the canvas with sensuous and "impressionistic" details. Something very similar could be said for the German Anselm Feuerbach (1829–1880), the champion of the so-called Römer, who were German artists still fascinated by the myth of Rome, not its primitive simplicity but rather the sensuality of the warm, Mediterranean landscape.

In England, we have the Pre-Raphaelite brotherhood whose main exponents are born precisely in this intermediate generational cycle, at the midpoint along the path that leads from medieval, Romantic "historicism" to photographic "naturalism": William Hunt, born in 1827, Dante Gabriel Rossetti in 1828, and John Everett Millais in 1829. The very apt label they give themselves when they form a society in 1848 has been mentioned several times and used *ante litteram* in the present study due to its effectiveness and relevance, especially given that the symptoms of contemporaneity I have attempted to identify begin with the tendency to look back to a period "before Raphael." I have also pointed out that this looking back may not be sufficient, like the antibiotic regimen that fails to eradicate the bacteria it is designed to attack, thereby permitting its survival and rapid propagation as soon as the medicine is no longer administered, as we hope in vain that the illness has been conquered. In fact, "before Raphael" we could find meticulous, excessively detailed realism that is certainly less "modern" than the sensuality and airiness of Raphael's own work or the work of the great Venetians, but still too compromised by mimetic representation. This was the *caput mortuum* that still threatened the art of the Nazarenes. It reappears in the art of the Pre-Raphaelites about thirty years later. We can still salvage Hunt and Rossetti since they agree, though in different ways, on the primacy of the hard-edged "design," thereby opposing comfortable naturalism. This is not the case for Millais who, by contrast, becomes the master of seductive, alluring scenes bathed in apt tones and atmospheres. It is commonly said that, after his early experimentalism, he was swept up by the Victorian climate in England, which dominated the second half of the 1800s. However, in that environment, Millais' masterpieces are perfectly "aligned" with the best European paradigms, resonating fully with the works of several contemporaries mentioned here: Böcklin, Morelli, and Lega.

In conclusion, the historicism dominating the first half of the 1800s fits very well with the return of modernity. In fact, it entails a rejection of the heroic and extreme looking back that had been attempted as a first step in a process characterized by "primitive taste." That historicism, however, quickly closes the chapter of revival of the Dark Ages, the Romanesque, and the Gothic. Even the "hard" 1400s are soon considered an unattractive, unliveable frame of reference. Having removed the barriers, we link up again with the golden centuries of modernity and its coherent development in the 1600s and 1700s, along with its unbridled and unabashed naturalism. Armed with this flexible and stimulating instrument, the artists of this period set out to capture reality, the intensity of the moment, even if for the time being they are limited to a skirting manoeuvre in the face of that reality and they leave the task of expressing such qualities directly to the generation of 1840.

This strategy, which is apparent in the work of most artists from those years and in that of their colleagues, the writers of historical novels or composers of musical works, is powerfully endorsed as well by philosophers. These artists are all prepared to adopt the "historicist" paradigm in order to reconcile the idea of origins and primitivism, which are the domain of the imagination, with the unavoidable bypassing of that idea in favour of a final phase of development marked by the triumph of reason and consciousness – in fact of self-consciousness. A large space for more technical philosophical speculation is left to determine whether such a result is a product of pure logic or the more secular achievement of the physical-mathematical sciences.

An essential role is played in all this, of course, by Hegel who is born into a generation that includes some of the most interesting champions of pre-postmodernity, which is the dawn of the contemporary age described in the present study. His dates of birth and death are 1770 and 1831, which are not far from Turner's or, to stay in a context closer to his, those of the Schlegel brothers (August is born in 1767 and Friedrich in 1772), both writers and aestheticians. Even more interesting are other areas of chronological proximity in German philosophy itself. Schelling is born in 1775; Schopenhauer is quite a bit later (1788 as has been noted), but those born around 1775 for the most part exhibit a vivid interest in the sphere of inquiry that emphasizes energy. They are prepared to declare the primacy of this sphere, although they subsequently have to admit that, in some ways, this sphere must come to terms with the opposite sphere privileging constrained reason. But their clash, or the attempt to have them converge, has to be carried out without sacrifices on the part of the sphere that stresses energy, considered to be the more interesting and vital, as well as the more original of the two, to the point of giving the face to an entire epoch.

Hegel, instead, devises a strategy that enables him to "have his cake and eat it too" by revising an analogous, diachronic schema already skilfully used by his Italian predecessor Giambattista Vico. It consists of allowing each sphere to develop fully but in different historical phases, and through a process that concludes to the advantage of the sphere that stresses reason. Such a schema proves to be favourable to the spirit of modernity and irrevocably pushes its counterpart, the imaginative (aesthetic and artistic) realm, into irrelevance. In the course of history, the epoch of art must "die" to make room for the epoch of reason. It matters little if the reason Hegel has in mind is very different from the analytical reason that deconstructs and measures, as it was understood in the 1700s. This is merely a technical detail, as has been observed already. What is more, it is also a known fact that Hegel's dialectical schema, which is based on the primacy of the spirit and the ideal (the only one to which the good of existence is attributed), is "turned upside down" by the so-called Hegelian left, from which would emerge both Marxism and Positivism; they are in agreement, at least in their awareness, that the culture of the middle decades of the nineteenth century must come to terms with the industrial-mechanical revolution as well as the enormous developments in technology and in the physical-mathematical sciences. In all this, we have the full realization of the modern age, which silences the timid intimations of postmodernity, about which I have spoken to the point of creating the impression that these intimations never existed and, therefore, the contemporary age must begin anew, redefine its problems, draw its correspondences and homologies exclusively from the last two or three decades of the 1800s, and wait until electrotechnics gains a place in the sun. From this point of view, the year 1789, proposed by the handbooks as the beginning of the contemporary age, would be ineffectual and would not have any consequences. Here again we may find a glaring disconnect between "official" periodization based on surface events and real periodization supported by the events and processes of material culture.

THE START OF THE CONTEMPORARY AGE

It is certainly true that, beginning with the decade 1860–1870, things proceed in an orderly and effective manner. Technology finally takes important steps, as noted above (Pacinotti's ring armature and the installation of the first transoceanic cables) and "high" or symbolic culture does likewise, in accordance with the analogy of the swift circulation of a liquid from one vessel to another, an analogy that is the basis of my reflections. A great physicist and mathematician, Maxwell, gives us equations that map the electromagnetic field and a great painter, Cézanne, paints very strange works that would elude interpretation and would have to be attributed to the painter's troubled youth, if we did not have the courage to see them as homologous to the turning points in other sectors of culture. This is the same "courage" that inspired my book, *Arte contemporanea* [Contemporary art], to which I refer the reader for a more detailed study of that fateful, inaugural decade. Here, I limit myself to considering some key moments in that history and to commenting in general on their culturological impact.

Cézanne's paintings from that decade exhibit a free style consisting of many undulating lines that really have no equivalent in the art of any contemporary, the best of whom, Claude Monet, was taking "modern" representation to the highest level of "photographic" realism and, therefore, adhered to the principle of the perspective pyramid. In Cézanne, instead, we no longer have the visual pyramid and, by and large, there are no longer straight lines. The space is curved like a reflective, concave or convex surface. It is also an immense space bathed in shadow broken only by flashes of light that have a mysterious, endogenous source. The subject can be defined as Romantic but in the thematic sense since the artist is very close to his contemporary, Émile Zola, and like his fellow countryman, is influenced by the works of Gustave Flaubert. He is definitely not interested in the ossified classics and prefers the stories of the Old and the New Testament, Miguel de Cervantes's *Don Quixote*, or *The Temptation of Saint Anthony*, a mixed blessing for Flaubert. At any rate, this permits him to depict humanity evoked in a dreamlike state, without relying on detail or the laws of representation in precise spatial-temporal terms. This is a crowd of phantasms that the artist can summon like wisps of smoke from Aladdin's lamp and either expand or contract at will, like tremulous, vaporous smoke rings or like bursts of wave trains in a fluid (liquid or gaseous).

The young Cézanne travels from his native Aix-en-Provence to Paris where he meets the leading artist of the early 1860s, Edouard Manet, creator of two official "scandals": *Olympia* and *Le déjeuner sur l'herbe* [The Luncheon on the grass]. These are scandals from the point of view of the theme and the daring, controversial situations represented, but also because the "representation" itself does not yield completely to the "modern" paradigm. Indeed, it exhibits a daring synthesis and bold flattening of the figures. Manet is a very interesting artist, an intermediate figure between the great pioneers of the late 1700s and the arrival of the "new" artists, founders of contemporary art, led by Cézanne. However, Manet lives in times that are not ripe for the full development of the early features of the postmodern elements we find in his spatial conceptualization; so much so that, a little later, on the threshold of the year 1870, he comes into contact with the exhilarating but limiting work of the younger Monet, who would bind him to the Impressionist project, despite the fact that this project was tight-fitting and suffocating for Manet's art.

Therefore, the work of Cézanne, who cleverly "copies" Manet's two scandalous canvasses, is liberating in some way since it compels his elder to take his art out of its state of suspension between the present and the future, giving that art a forward-looking quality. This, however, means that we need to interpret the work differently. If Manet flattens and stretches the bodies, Cézanne comes along to endow them with an electrical charge that makes the figures seem to dart, become compressed, fold in on themselves, only to stretch out again in alternating cycles.

But for Cézanne that project of adapting slim, streamlined bodies to a spherical space is too advanced. A short time later, beyond the threshold of 1870, he too is attracted by the triumphal march of Impressionism, which he encounters especially in the person of Camille Pissarro. In a sense, this is a situation we already know. The good intentions of artists who set off on the right foot, determined to give us slim, synthetic, courageously distorted images in keeping with the poetics of curved lines, are betrayed by a return of the modern preoccupation with detail and hard-edged images. As a first step, the Impressionists reject the possibility of both images produced in the studio and "mental" compositions. The easel must be taken *en plein air* or outdoors in direct contact with nature, in accordance with the "modern" ideology and its fascination with the physical-mathematical sciences, which are grounded in experimentation. Cézanne is somewhat embarrassed by his "literary" past and adheres to the new work ethic displayed by his Impressionist associates; thus, he risks proceeding to "normalize" his vision. Fortunately for him and for us, his intuitive genius resists this outcome and limits him to making only some corrections to the spherical space. In fact such a spherical pattern is suitable only for schematic or essential forms, whereas it does not satisfy the requirements of analytic representation. However, all the artist has to do is make a few appropriate modifications, such as paring it down and imposing upon it many regular facets. The result is the polyhedron, which does not negate in absolute terms the principle of the sphere. (Even in modern geometry the sphere is described as the point of arrival, the limit of a regular polyhedron, where a number *n* of facets tends to infinity). But, each facet can receive a "quantum" of physical data, present a surface on which to concentrate colour and light, and it can contribute to the construction of the image as a whole. In this way, a mature and definitive style of the artist from Provence emerges during the 1870s, a style he would not abandon for the rest of his life. It consists of using a polyhedron, similar to the "fly eye," frosted glass, and crystal that filters the image through its facets. In general, the primacy of the new, curvilinear style suited to the technotronic age is confirmed. At the same time, it takes a step in the direction of a compromise with the preceding mechanical, analytical Cartesian age by adopting a compositional unit that is very "normal" since it still respects the principle of modular-rectangular forms. In this, Cézanne anticipates the long battle, conducted in the subsequent stages of development in contemporary art, between electromorphism and mechanomorphism, where the former tends to prevail, but not without significant comebacks and revivals on the part of the latter.

Despite the power and compelling quality of his works, Cézanne remains an exceptional figure, "outside the chronological window" since he is born in 1839. This makes him a contemporary of the Impressionists who are active both in France and in other parts of the western world. As I have just noted, he certainly does not reject *en plein air* subjects or airy and detailed landscapes even if, in reality, he anticipates the radical forms that contemporary art would soon produce. In this sense, Cézanne is an artist "outside the generational window" just as Leonardo was in his way a few centuries earlier, although their respective talents took them in opposite directions. Leonardo was exceptional and outside his chronological window anticipating the mature "modern" style that would be fully realized by the masters of the early part of the 1500s. Though he seems to pay homage to the principles of Impressionism, Cézanne produces the most arduous "contemporary" canvasses.

This generational anomaly disappears if we consider briefly the members of the next wave of artists who belong to the "generation of 1860" that gives rise to Symbolism, which, from a technical-visual standpoint, would be articulated into more precise "isms" like Synthetism, Cloisonnism, and Divisionism. As for more properly linear components of the composition, we can speak of phytomorphism (or floralism). One of the founding fathers of such a cast of mind, with whom the contemporary age really begins, is once again someone who "precedes" his generation. Paul Gauguin is born in 1848 and, therefore, in the immediate rear-guard of Impressionism; in fact he succeeds in associating with these artists thanks to the always useful mediation of Pissarro (who also assisted Degas and Cézanne). Gauguin, however, definitely "slows down" and reduces his reliance on detail or the "impression," the core of his older contemporaries, since he was propelled by a frantic search for synthesis and reduction. This was for him a long and difficult quest precisely because he was born much later than the others in the group – so much so that the contribution of some young artists who were more in line with the average date of birth would be indispensable for him. Two of these artists are Louis Anquetin and Émile Bernard, who established their courageous and innovative "ism," Synthetism, in 1886.

Lastly, the prophecies made less than a century earlier by Blake but also by Flaxman, Goya, and all the *homines növi* [new men] about whom I have spoken, finally come true. Contemporary art rediscovers its vocation for abstract forms (in the etymological sense of the term, whereby the essential is "drawn out" in stringy shapes onto the surface of the paper or the canvass). The contour lines of these "cut-outs" or *cloisons* are flowing and curvilinear in homage to the wave principle that underpins the entire technotronic era; they could also be seen as phytomorphs. In the last part of the century, against the rigidity of machines, right angles and their derivatives, artists choose a more accessible and more familiar principle, that of *natura naturans* [nature nurturing], as it appears in the growth of vegetation (plants, in Greek *tà phytà*). This is why we have to refer to a general phytomorphism typical of the forms produced in that period (which consistently relied on vegetation, springtime, and blossoming plants: *Ver Sacrum*, *Evergreen*, *Jugendstil*, etc.). In hindsight, we can say that even then they sensed the need for electromorphism, but it was realized through the mediation of nature, which is closer to the new sources of energy than were the forms of the industrial-mechanical world.

Paul Gauguin is not the only leading figure in that setting, which includes all the groups he himself established, such as the Pont-Aven School and the Nabis. Still within the Paris avant-garde, he may be compared to the younger Georges Seurat, whose date of birth (1859) is perfectly aligned with that of many kindred spirits scattered in various places of the west (James Ensor, Edvard Munch, Gustav Klimt, Ferdinand Hodler, as well as the older Italians Gaetano Previati and Giovanni Segantini). At first sight, it appears that the two choose opposite paths: Gauguin proposes to paint with broad brushstrokes, while Seurat advises using shorter strokes, though not necessarily reduced to a point. The final result, however, is for the most part a convergence of the two approaches just as it is possible to reach the East Indies by traveling eastward or, less easily, westward. In the end, the decision as to which route to take becomes a question of expediency, economy, and calculation of the advantages and disadvantages.

At any rate, a new practice is born, that of compiling "discrete" images made from discontinuous elements, images that, for this reason, are fundamentally the opposite of those in the photographic-impressionistic style, the final phase of the "modern." These last images seem to be based on a register of nuances and gradations. In these images, owing to the laws of optics, it becomes completely natural for bodies to have no limits, no linear boundaries, and to dissolve into the atmosphere. Leonardo had intuited this and in so doing won eternal fame for himself as the "forerunner" of modernity. The Impressionists had confirmed as much barely 10 or 15 years earlier, by performing a gesture that seemed courageous, blasphemous, and revolutionary; that is to say, they accepted at a "high" culture level, what a "low" technological medium, photography, was producing in great numbers.

But at that very moment, "high" culture was quick to reject the normality of continuous, detailed, and smoky images. Of course, any innovation has to have some correspondence to technological advances, according to the methodology used in the present study. In this case, the nascent advertising industry was demanding the production of "thin" and synthetic images entrusted to the essential style of the affiche, or ad. It is not true that the age in which we live conditions people to consume high-definition visual messages, like those provided by photography and later cinema. As a minimum, we have the parallel development of a series of "thin," abstract means of reproduction that use old techniques called back into service: etching and lithography. What is more, we have a series of technological innovations that confirm decisively the primacy of the "discrete." In fact, today's society does not so much use photographs (which remain the individual act of a craftsman) as their transcription onto plates ready for the printing press or for long-distance transmission. In other words, here we have the logic of the lithographic "screen," a conversion device that "reads" the continuum of "modern" images (photographic and impressionistic) in abstract terms, extending exponentially the Divisionist technique of Seurat and his associates. Divisionism, however, was not an isolated moment of idle speculation on the part of avant-garde groups subject to the usual accusations of snobbery and elitism. At least, they acted like true "avant-gardes," anticipating or sampling in advance mass-consumer devices that would inevitably come after them.

Today's visual technology, including the photolithographic screen, a process of telephotographic transmission, and even the "electronic mosaic," which leads to television thanks to the invention of the cathode tube, and so forth, confirms the validity of the anticipatory insights of the Gauguin-Seurat, Synthetism-Divisionism pairings. The process is not yet finished. There is another important fact. If they were harbingers of mass technology, the projects of the avant-gardes also reintroduced old techniques that "modernity" had discarded and tossed into a museum of antiquated practices. The segmented and discontinuous image had already been produced over long periods of time in different cultures and it too deviated from the naturalistic-illusionistic style of previous eras. Consider, for example, the Byzantine mosaics in Ravenna, Gothic stained-glass windows and tapestries, which were made for many centuries, but always with their roots in pre-modernity. In sum, the modern age completes the great circle uniting the past to the present and the future.

OPTIONS IN THE EARLY TWENTIETH CENTURY

The lines of development in contemporary art are many and far from univocal. The end of the 1800s is dominated by flat, stylized, "abstract" (during the Symbolist period) images, and this predominance seems to continue into the first decade of the 1900s in the works of the Fauves and Expressionists. But at a certain moment Cubism emerges, a style that many would agree to consider as one of the most important and substantial episodes of the contemporary era. This "ism," pioneered by Pablo Picasso (1881–1973) and Georges Braque (1882–1963) as well as others, crosses a different threshold than that of abstraction to arrive at the "concrete" and the creation of autonomous, self-grounded forms that no longer rely on nature. "Abstract" artists, still conditioned by the legacy of the modern-impressionist era, limited themselves to working on the contours of natural objects from the exterior, schematizing and reducing them, but somehow acknowledging their primacy. The "concrete" approach, instead, does not entail intervening on nature, but replacing it. Human beings develop total faith in their own capacities and think they can create objects, emulate nature, indeed, improve it, and produce objects that correspond better to cultural-technological exigencies and to the models of rationality employed in other areas of operation. This opens the path to establishing, without having recourse to mediations or lags, an explicit technomorphism whereby art gives complete and autonomous form to processes on which the technology of the same historical period is based. To some extent, this goal eluded the Symbolist avant-garde, which was already able to intuit the new electromagnetic energies, but was unable to give them explicit visual form. As a result, the avant-gardes were forced to pass through the filter of the plant-form (hence the phytomorphism discussed above).

The Cubist phase of Picasso and Braque's technomorphism (1907–1914) represents a giant step forward because technology is given direct visual representation, but there is also what I would not call a step back but a deviation or pause since Picasso and Braque make a diagnosis that is only half-right for the times, with its limited capacity for predicting the future. In fact, they "bet" on the inevitable and lasting primacy of mechanical technology; for them, "machines" are a sort of unavoidable *a priori* of all human activity and so trees, bodies, and faces have to be reconstructed according to these "mechanical" parameters. Nature is "defective" or at least it is guilty of wasting energy precisely because it habitually uses unusual and complicated curves to produce its objects. In the twentieth century, technology, art, and human culture have to do more and they have to do it better. The modules of artificial production need to go outside their own areas and be applied in all possible acts of creation. This is the task taken on by the Cubists.

In practice, they reach the point where Cézanne had arrived. We have seen that the master from Provence started out with a spheroidal model that was already electromorphic to a large extent. He "modified" that model, revising it through the use of rectangular units, which are more reflective of "modern" taste. But in Cézanne's case, it was not clear whether these units belonged to *homo faber* in a redesign and reconstruction phase or to "natural" man who finds them in the perceptual field provided by Mother Nature as an *a priori* of sight, as opposed to inventing them. In this way, Cézanne did not belie his stylistic ties with the Impressionists on the basis of direct contact with the "subject." He too was not reluctant to go outdoors to execute his paintings, although he would later work patiently on the image in his studio and for long periods of time.

In the case of Picasso and Braque, "going to the subject" loses all meaning because it is no longer a matter of "perceiving" nature, even with essential templates, but rather "remaking" it the way architects or engineers design models of buildings or machines in the studio or on the drawing board before proceeding to make the object. Determining whether this process produces bodies and objects that can compete with those of nature or, instead, is fully satisfied with its autonomy is less important. In other words, Cubism can also be "figurative" in the results it produces, as it has been in each of its phases (Proto-Cubism, Analytical Cubism, and Synthetic Cubism) without diminishing the legibility of its subjects. But the internal logic of those images challenges nature and attempts to replace it.

In this case, it is inaccurate to speak of images, which is a convenient term as long as the artist intervenes on nature, reducing and transferring it from its inherent three-dimensionality to the two-dimensionality of the surface medium, which visual art has a long tradition of using. But in an *ab imis* [from the deepest foundations] reproduction, the image is only a secondary phase. Even architects use it and perhaps in this case they are using all the tricks of the tradition to suggest a real object (dark shading, axonometric projection). The most pertinent outcome of this process, however, is the real object in its plastic concreteness, an object among other objects, a heavy and solid thing. This is also true for the moment of painting in Cubist art, which, though splendid, must be considered transitory and anything but definitive. True Cubists should attain three-dimensionality and plasticity and give material existence to the "cubes" they sketch, taking them out of the dead end that is the fiction of surface.

In effect, Picasso, who remains the most inventive of the Cubists, took that very step on several occasions by giving us plastic aggregates of objects and relief prints that project from the surface even if they do not give the impression of becoming completely detached from it.

In other words, the threshold of concreteness leads to a series of processes that are increasingly bold and cause us to surpass the logic of fiction used in art for millennia, but taken to the extreme or perfected in the modern-photographic epoch, with its capacity to offer high-resolution mirrors of reality. Now the concept of the mirror wanes. Humans "know" by sending into the atmosphere beams of concentric waves that quickly envelope objects, constantly changing the point of view and superseding that logic by substituting it with another in which "the centre is everywhere" and three-dimensional objects are physically present, ready to be manipulated. The *perspectiva*, or seeing from a distance, is eliminated completely and no longer needs to be replaced with equally illusory systems because it is the very principle of long-distance simulation that is eliminated. On the other hand, a similarly ubiquitous presence of three-dimensional objects, owing to the possibility of enveloping them with wave trains, acquires a dematerialized quality. In a sense, it is difficult to say whether the objects themselves are physically perceived or are instead "thought" or evoked almost magically. This also falls in with the processes designated by the term "the death of art," understood as the end of the technical, artisanal principles that, in the past, led painters and sculptors to produce images with the aid of graphic signs or patches of colour applied to a physical surface. Now artists propose mental images; they "think" objects and situations without having to produce them materially since, from this point of view, they find these objects everywhere already; it is no longer a matter of adding one object to another, but rather of making healthy and therapeutic use of what already exists.

By mentioning these processes, I have gone beyond the experimentalism of the Cubists, despite the unquestioned boldness of that experimentalism, which arrived, as it did for the restless Picasso, with the practice of collage: objects or fabrics (oilcloths, newsprint) taken from life and, therefore, beyond the threshold of representation, even if for the moment juxtaposed to the most conventional pictorial forms. As I have pointed out, in his bolder moments, Picasso made use of projecting "objects," made in relief but always situated in an ambiguous area in relation to traditional practices.

That use of real objects was destined to become more courageous in short order. There was a general trend to move from simple collage to assemblage, wherein "real" or "concrete" objects became increasingly important. This occurred with the cluster of movements associated with Dadaism. One of these is represented by German Kurt Schwitters who emphasized the use of "found objects," selected because they have a high degree of sensuousness and provocativeness, even if pertaining to the ugly, the unpleasant, and the off-putting: objects produced in the factory, objects that are now worn, forgotten, and faded by time and the ravages of the atmosphere.

There was, however, a more disquieting and radical strand of Dadaism centred around the figure of Marcel Duchamp (1887–1968), who took the logic of using "ready-mades" to its extreme by attributing its provocative power to the fact that these objects were not selected for their aggressiveness but for their passivity and anaesthetic neutrality, which is to say, they lack the capacity for sensory stimulation, both positive and negative. Having become "mental operators," artists apply themselves to the task of affirming the aesthetic properties of every object and situation that pre-exists them. All they need to do is want it, declare it, and apply an appropriate "intention." These intentions must be directed not only at solid, material objects. They can just as easily be directed at verbal utterances, atmospheric phenomena, or the products of mental reasoning. The whole universe can be rethought in terms of its aesthetic properties if we apply to it appropriate coefficients (like pressing a key that causes all the characters on a page to be typed in capital letters).

Taking a step back with respect to Duchamp's very bold practices, which were isolated cases in the first part of the twentieth century and had to wait until the second half of the century to see their full development, and returning to the domain of virtual images, I must mention other important movements, in addition to the mechanomorphism of Cubism and affiliated movements. These include the Dutch Neoplasticism of Piet Mondrian and his followers, the Suprematism of Russian artist Kasimir Malevich, and Constructivism, an important trend in Revolutionary Russia (1917) led by sculptor Vladimir Tatlin and his followers. Wassily Kandinsky (1866–1944), another famous Russian artist, developed an almost opposite project with which he arrived at the "concrete" himself (though he always preferred to talk about "abstraction"). Around 1910, he pierced that "veil of Maya," the delicate screen on which, up to that point in his career, he had projected images in the style of the Fauves. Once the veil was removed, the fullness of being revealed itself, understood as a sort of amniotic fluid or cellular fluid seen under the microscope where life unfolds in all its dynamism and creativity. From this emerges the concept of radical biomorphism where art, with all its traditional instruments including graphics and vivid colours, simulates the "open" and unpredictable morphology of embryos and viruses. This is still a journey through nature, but this time it is a profound nature, a *natura naturans* [nature nurturing] that abandons bumpy surface features and exhibits strict affinities with electromorphism, even if the fusion of the two fields is not permitted.

On the contrary, authentically electromorphic works can be found in Italian Futurism, one of the most important movements of the early 1900s, and especially in one of its most prolific exponents, Umberto Boccioni (1882–1916). In terms of the crucial "thresholds" identified in the present study, his brief but stormy career can be summarized as follows. He starts with fin-de-siècle flat, stylized forms, which he encounters in the versions provided by the great Central-European exponents of Symbolism (James Ensor, Edvard Munch, Gustav Klimt, and Gaetano Previati), but he exaggerates these forms in the direction of Expressionism. He is on the verge of freeing himself from the *caput mortuum* [worthless remains] of subordination to Nature, thereby producing a "concrete" Expressionism, consisting of ample puffs and rolls that spiral into space. However, along this difficult path to the concrete, in 1911, Boccioni is influenced by the mechanomorphic "concreteness" already developed by the French (not just Picasso and Braque, but Fernand Léger, Robert Delaunay, Albert Gleizes, and Jean Metzinger as well). He then begins to contain his vitalistic impulses within a cage, but at the same time he does not refrain from endowing them with spatiality. In other words, in Boccioni's post-1912 paintings, the two groups of geometric patterns, one consisting of solid bodies formed by hard-angled dihedrals (cubes and parallelepipeds) and the other consisting of bodies formed by the rotation of eccentric curves (parabolas and ellipses), coexist in a relationship that is somewhat incongruous but important because it illustrates the inevitable coexistence of the two great technological families of the last century: the mechanical and the electromagnetic. Boccioni also has the audacity to extend this difficult morphological coexistence into the domain of plastic forms, i.e., sculpture.

Finally, another Italian "signs" the last truly decisive epochal project that should be included in the crowded early 1900s and is worth discussing in the present study, which, let us keep in mind, is not a "history" but a theoretical reflection. I am referring to Giorgio De Chirico (1888–1978) who, from the time he begins to paint, consistently rejects the poetics proposed by his more brilliant avant-garde contemporaries in various parts of Europe. In short, he does not believe in the present or the future; he does not believe in the possibility of developing either the virtual images of the "machine age" or those of electricity and biology, nor does he believe in a "real" use of objects and situations that are immediately present. But he does adopt a technique that is very close to Duchamp's "ready-mades," even though with a radical difference. Basically, he invites us to consider that today's artists do not have to wrack their brains inventing new styles since we have had perhaps too many styles in the past and, therefore, all one has to do is revisit or cite them the way Duchamp does with found objects that are literally present in his work. In sum, the "ready-made" technique needs to be redirected toward past circumstances evoked in their outmoded fashion, indeed their revolting traditionalism, a technique that seems to oppose the poetics of the avant-gardes. Naturally, "citations" must also introduce "distorting" elements, even if these serve only to lighten, suspend, or accelerate the process. Therefore, the objects De Chirico appropriates from the rooms of an ideal museum of the history of art and architecture overlap and short-circuit one another. This can also be equated to the so-called dream work, which is an area of experience in which we evoke the perceptual information in our database, but in spontaneous and unpredictable ways. The dream work only anticipates what is made possible today by computer databases or, more generally, by the enormous quantity of accumulated knowledge typical of our advanced civilization.

De Chirico, for the most part, and after him other Italian artists who practised Metaphysical Painting (Carlo Carrà and Giorgio Morandi), exhibit the powerful relevance of an approach based systematically on revivalism (the recovery and recycling of forms and styles from the past). But all this was already anticipated by the innovators of the late 1700s who, as a first gesture, refused to sanction the flow of modernity as it made its way more or less serenely from the time of Raphael onward. It was time to disrupt that continuity by introducing a historicism interrupted at intervals by long, deep fissures and by plunging deeply into the past.

NORMALIZING PROCESSES OF THE SECOND HALF OF THE 1900S

It is difficult to argue that there is something decisively new in the second half of the twentieth century. However, the ferment, theories, and projects that appeared in the early part of the century undergo tremendous growth in quality and especially in quantity. Briefly, we can say that the second half of the 1900s "normalizes" the insights of the historical avant-gardes, drawing them out of their aura of revolutionary and solitary uniqueness to give them a more generalized status.

The Second World War brings about a collapse of faith in the "machine age" because machines turned out to be harmful and spread sorrow and destruction, and because a new, nuclear technology finally emerged. At first, it too aligns itself with more traditional technologies that cause death and destruction. As a response, an anti-technological mindset develops that, if anything, attempts to create an alliance with the logic of "the living world," that is, with the organic forms of the biosphere. At that moment it is as if early twentieth-century biomorphic inquiry got its revenge on mechanomorphic inquiry, which was dominant in the past. Wassily Kandinsky comes to the fore and perhaps more than he (since he is guilty of having surrendered to a certain extent to the constructivism promoted by Walter Gropius and the Bauhaus school in the 1920s) the masters of Surrealism triumph, especially those of the fluid-organic variety: Joan Miró, Henri Masson, and to a lesser extent, Max Ernst and Yves Tanguy. Therefore, we have the blossoming of styles called Informal, Tachisme, and Abstract Expressionism, in a substantial meeting of the minds on both sides of the Atlantic, with contributions now also coming from the "Pacific cultures" and in particular Japan, which recovered rapidly from its military defeat. One could observe that all this climate, characterized by unchecked vitalism, does not provide any support for my theory of a verifiable homology between the stylistic choices made in the field of art and advances in technology. It has been stated above, however, that it is possible to establish a good correspondence, if not an outright homology, between electromorphic and biomorphic products. At a minimum, the climate of the Informal does not contradict the "soft," curvilinear quality of electromorphism. At any rate, it reiterates its aversion to designs that are too rigid and orderly. What we have, instead, is a certain respect for typically "modern" interpretations of the act of painting, which continues to be performed on a flat support (canvas, for the most part) and with paints. The final location of the object produced (the "painting") is the traditional home or art gallery "wall." Nevertheless, in many of these environments, the Informal already tends to take innovative steps. For example, the coloured materials often appear to be thick and may be taken directly from daily life, for instance, the tar used by Jean Dubuffet. The techniques for applying these materials are also irregular (palette knives, sometimes the hands of the artist, whose body begins to assert its presence; the artist is no longer satisfied to remain hidden behind his or her implements).

But the second half of the twentieth century witnesses the triumph of the practice of manipulating objects directly, "presenting" them, rather than "representing" them. This is the end of the age of the mirror image executed on a flat support since the new technologies based on electromagnetic waves are incompatible with such procedures. At first, in the 1950s and 1960s, there is a period during which the inside and the outside of the painting surface coexist. Typical of this phase are the "combines" produced by American Neo-Dadaists, Robert Rauschenberg and Jasper Johns. The label with which they are designated is also telling since it indicates that Dadaism appears as the most incisive of the historical movements, even if for the moment the two American artists do not know how to (nor do they want to) acknowledge the many forms of historical Dadaism, especially the distinction between Kurt Schwitters' loud, picturesque version and the more subtle and systematic version of Marcel Duchamp. They offer us the direct application of the object, however, enlivened by painterly-manual interventions. Moreover, the object is limited to projecting from the picture plane without attaining full autonomy. Similar features can be found in an analogous movement born in Paris, that movement being the Nouveau Realisme [New realism].

The early 1960s appear to signal a return of the industrial-mechanical society, which attains full maturity during that time. These are the co-called economic boom years, which flood the markets of the western world with cheap merchandise in a substantial "popular" levelling of taste as well as of the standard of the objects themselves.

This brings about an ephemeral return to "representational" techniques in the cluster of styles that take the name of the most prominent of these, Pop Art. Mass-produced "popular" and common objects are celebrated, but through the same filters used by mass media (for example, the photolithographic screen, the blow-up, and comic strip stylization), which also produces a resurgence of the climate that already existed during the time of Symbolism with Gauguin and Seurat. However, mass-produced objects are not always "represented." The notion of "presenting" objects in three-dimensional forms or reproducing objects to the point where they compete with the very mechanisms of industrial production also continues to develop. These objects are distorted and exaggerated from the moment they are produced, in accordance with the requirements of artistic-aesthetic intervention. In fact, American Pop Art spawns the "flat surface" artworks of Roy Lichtenstein and Andy Warhol as well as the voluminous sculptures of Claes Oldenburg, along with many intermediate styles where the object can break through the surface or be absorbed by it in a lively process of receding and emerging (Jim Dine, Tom Wesselmann, James Rosenquist).

Finally, 1968 marks another turning point. Perhaps it indicates the final defeat of the "machine age'" and the opulent society that produces consumer goods in large quantities but sacrifices quality, resulting in the levelling phenomenon of forced urbanization. Conversely, we become aware that the technotronic society has now reached full maturity, but it demands that all the exponents of the avant-gardes be in synchrony with its wavelength. The culture of the object (whether offered as representation or as direct presence) comes to an end while interest shifts to processes, actions, and behaviours. This also means a return to the Informal, the prevalence of vitalistic tendencies based on the primacy of the flow of energy. But while the Informal of the 1950s was still tainted with residue of the "modern," due to the fact that it remained closed within the picture frame and continued to rely on the rituals of painting, this new Informal puts its interventions in the open air and proceeds to capture the immediacy of events. The artist "works" directly on spaces, which can be those designated by art galleries, but they can also be "found" in daily life (urban areas, suburbs, and deserts) without forgetting that, beyond the physical dimension, which has been expanded beyond all proportion since the traditional limits of the picture frame no longer exist, it is now possible to plumb or "massage" the invisible, intangible dimension of thought and ideas. In fact, the new post-1968 frontiers are located simultaneously within an extreme materialization (nothing is sufficiently large for the artistic gesture) and an equally bold dematerialization whereby the artist limits himself or herself to minimal use of the physical supports provided by numbers and letters; that is to say the artist reduces the "signifiers," as they would say in linguistics, to a minimum and invites us to explore especially the realm of the "signifieds," pure noetic phenomena.

There are many ways to describe further the particular climate created after 1968. One of these is to resort to McLuhan's notion of the "cool" medium as opposed to the "hot" medium. For example, the historical Informal of the 1950s would be considered "hot" because it was still reliant on the primacy of vision, offering us works that first had to be seen with the picturesque design of their tumultuous subjects. To this was added a psychological meaning: those paintings were hot because they contained the intense pathos of the artist.

The new Informal of the late 1960s, instead, would be "cool" since it seeks to establish a correlation among the sense organs. In the works executed with this technique, there is not much to see, or at least the visual data has to be supplemented with acoustic elements (there is often a tape recorder or recordings that are an integral part of the artwork), and especially haptic data, that is, elements that stimulate our sense of touch and our experience of the three-dimensional world. Finally, not to be overlooked is movability, which is the fact that the artist or someone acting on behalf of the artist, causes the installation to function or invites spectators to participate, thereby ending their role as passive and external contemplators.

But having seen these aspects of the procedures performed, the question arises whether this should still be called art, which, according to the Latin root word, is the idea of producing objects using refined technical skill, or rather aesthetic activity, designed to excite and intensify our aesthesis, i.e., our capacity to experience sensations. There is a good deal of debate on the issue of "the death of art" and the shift to the intense promotion of basic sensibility. We are also encouraged to replace the term artist with "aesthetic operator."

We are still dealing with a cluster of very interesting styles distinguished by many labels that do not, however, hide their essential unity, especially if they are seen through the culturologist's penchant for amalgamation: Minimalism, Anti-Form, Process Art, Land Art, Body Art, Fluxus, Behavioural Art, Concept Art, Arte Povera [Poor Art], etc. Certainly, there is a perfect homological equivalent for technotronics in this climate, which dominated the early 1970s. Sometimes there is a direct application of its discoveries and not only a homology with a "mental" level, which can be demonstrated only through abstract argumentation. As a matter of fact, the "aesthetic operators" of that entire period do not hesitate to use explicitly the technical discoveries of neon, refrigeration, electrical resistance, video recording, and so forth.

But we should not fail to mention a counterstroke that is very similar to the one produced in the second decade of the century when the explosive phase of the styles led by Cubism was followed by the implosive phase led by Italian Metaphysical Painting. Once again, the second half of the twentieth century, through its contradictions, confirms the general characteristics to extend, normalize, or "cool" in every sense of the word, from the material to the psychological.

Indeed, some practitioners of conceptual art decide to apply the instruments of that art (numbers, indices, photos, etc.) not to daily occurrences, the "here and now," or the future, but to the "there and then," i.e., the past and more or less remote history. We are, therefore, faced with a new historicism and revivalism, which was already well initiated by late eighteenth-century artists, as noted above. Therefore, we have not avoided the basic options that the postmodern era offered from its inception, confirming the inevitability of a return to that fateful date of 1789, as the handbooks propose, even in an unacceptable manner, in order to find a reliable border between the modern and the postmodern.

Naturally, it is not that the various phases of revivalism or "citationism" always reappear in the same form. Rather, there are substantial differences between them; for example, the most recent phase, which we experienced during the 1970s, was characterized by a maximum degree of "coolness" since artists ("aesthetic operators") who lived it relied on a variety of resources, many of which are "mental" (letters and numbers) or technological. This does not mean that "hot" instruments of painting cannot reappear in such a context. In fact, the 1980s witnessed the return to a pictorialism that can sometimes be exaggerated and is, at any rate, "hot" in the various senses of the word, even though a new period of "cooling" ensued. Will this century be characterized by alternating "hot" and "cool," explosive and implosive phenomena, or will a technological innovation emerge capable of leading us into a completely altered world? I have no doubt that, in such a scenario, the arts will play their role as intrepid precursors without waiting for their course to be "determined" by pressure from other fields or other areas of cultural production.

6

Heinrich Wölfflin and the Phenomenology of Styles

THE PRIMACY OF GENERAL FORMS

Heinrich Wölfflin was born in Winterthur, Switzerland in 1864, and therefore is a contemporary, give or take a year, of Husserl, Bergson, Saussure, and Dewey;[1] hence he belongs in the generation of great cultural "operators" whose historical task was to assert the priority of the general over the particular or the phenomenal. Each in his own way, these protagonists affirmed that we have an innate tendency to generalize and that this tendency does not "come after;" it is not secondary and derived relative to the capacity to perceive individual cases. In order to claim the primacy of the givenness of general forms, these figures had to engage in a courageous struggle against the various forms and strands of the positivist mindset, in tune with the classic statement of all nominalists for whom general occurrences emerge *post rem*. In nominalist belief, prior to ideas, there are material elements, atoms of sensation, particulars of phenomenal givenness; then, by abstraction we derive *simulacra*, fixed approximations that can never aspire to dignity of the first order since they will always be considered secondary products, like extrapolations made for the sake of convenience. For our protagonists, conversely, general forms are per-

1 Saussure was born in 1857 and the three philosophers in 1859. For a comment on the statistical significance of this coincidence of dates of birth among members of the same cultural epoch see Chapter 3.

ceived *in re*;[2] they are seen in the very body of phenomena. The possibility of distinguishing the individual aspect from the general or collective is the result of very complex operations. If we really wish to assert a priority, this priority would pertain to the generalizing instinct, but not because it comes before in terms of logic or chronology; in other words, it cannot act *ante rem*. The two processes unfold on equal terms, one for the other; therefore, we will not find in these authors a mindset that leads back to Platonism and the reality of essences. General forms appear only when we devote ourselves to studying the existing material, that is to say the vast phenomenal field. Therefore, for the various "operators" mentioned above, there is a legacy, conscious or not, derived from typical Kantian[3] thought, according to which the forms of sensibility and those of the intellect are not conditioned by experience, but otherwise come into play only facing experience. They are "already there" when we engage with experience without the need to ask what their origin is or to attempt to examine them in and of themselves.

It is worth noting that the battle of all generalizing mindsets like Wölfflin's is waged not only against the types and families of nominalists (empiricists old and new, positivists), but also against a spiritual family of a completely different origin, yet one that has also come to the fierce defence of particulars: Hegelian idealists, especially the authoritative strand and the last school represented by Benedetto Croce who, after all, was himself a contemporary of Wölfflin.[4] Moreover, we know that there was a certain compatibility between the two, who were brought together by the project they shared with an entire generation, that of opposing positivist atomism. They went their sep-

2 I am applying retroactively comments that Luciano Anceschi normally makes each time he introduces and defines the notions of poetics and literary institutions. See, for example, *Le istituzioni della poesia* [The institutions of poetry] (Milan: Bompiani, 1968).

3 In fact, I discuss their Kantian derivation, understood in a sufficiently broad manner, in *Tra presenza e assenza*, [Between presence and absence] (Milan: Bompiani, 1981).

4 He was born in 1866. Croce's main comment on Wölfflin is contained in the chapter, "La critica e la storia delle arti figurative" [Criticism and the history of the figurative arts], in *Nuovi saggi di estetica* [New essays on aesthetics] (Bari: Laterza, 1958).

arate ways, however, because Croce's philosophy was aimed at reconciling the particular and the universal and finally privileging the former. In Croce's aesthetics, there is a whole host of factors to account for a successful artwork, but these are filtered through an individual quality that impresses itself on these factors in an unrepeatable fashion. As a result, it is not possible to write its history or to attempt a comparison with other works of art from the same time and space. Each is an individual work. It may be possible to write its history or attempt a general comparison, but in that scenario we would have to settle for pragmatic *a posteriori* categories. It is worthwhile to point out that Croce found himself in the same situation as the unpopular nominalists-positivists. For him too, art, literature, and poetry can be reduced to a series of monads that do not communicate with one another. In the presence of these monads, the only thing we can do is recognize them and acknowledge their existence or attempt to produce rigorously custom-made descriptions and interpretations.

The clash of opposites and the tension between a positivist mindset and an idealist one, however, were so fierce and widespread for decades that even today experts and non-experts alike argue on the basis of these two equal and opposite biases, which lead them to reject "generalizations," especially in the fields of art and aesthetics. They insist that we have to respect the particularity of the great works and not dare defile or insult them with comparisons and assimilations. These are always impertinent, even in the most current sense of the word, that is, offensive to the dignity of the masterpiece. Only the icy intellectualism of critics dares to do the impossible, that is to say, plunge the surgeon's scalpel into the individual bodies of art and poetry in an attempt to perform their deplorable clinical studies or to carry out vacuous classifications. Fortunately, the density, the almost carnal organicity of great single works of poetry, repels these attacks on their marvellous wholeness.

Complicating the mental habit of Wölfflin and his generational peers is the fact that they did not limit themselves to identifying general forms; they also described their changeability. They drifted away from the teachings of Kant, to which I suggested we can connect part of their attitude. In the Kantian system, the forms and categories are given once and for all; they are outside time in an atemporal, static state. By contrast, for Wölfflin (but also for Husserl and for Saussure, in which case we need only think of "languages" as general forms), the great families of collective forms or styles change across time; they succeed one another. Those inspired by the methodologies of these authors, have two tasks to perform: one static and the other dynamic. They are obligated to carry out two kinds of inquiry. The first is to relate the phenomena to the general forms, that is, identify the link between the individual work and the style of an epoch; but they also have to observe how, at the same time, the great stylistic families change over decades, how one follows the other with its distinctive features. In short, it is a matter of relating the individual phenomenon to the type (the style or language) and then examining how these types succeed one another or how they change in the flow of history. This obliges us to formulate a law of variation or a typology of changes, in other words, a phenomenology of styles.

With respect to the first task (relating the individual to the type, family, or style), the juxtaposition of Wölfflin and Saussure[5] I have repeatedly proposed (on the basis of generational proximity) is largely a question of the relationship between *langue* and *parole*. We know from the statements of the founder of contemporary linguistics that this is a lively and dynamic relationship. We do not have to sacrifice any of the intensity or uniqueness of the single act (linguistic, artistic, moral, or political: the relationship between the two entities appears in every sphere of action). It is sufficient to have a firm understanding of the fact that all speech acts acquire meaning only against the backdrop of the rules of the language. They are two superimposed slides that enliven and integrate each other. The general indicates the outlines of the phenomenon, which on the other hand are concretized and either confirmed or undermined through acts of deviation that, up to a certain point, can be contained and reabsorbed. If they go beyond a certain limit, however, the deviations proceed until they ensure the foundations for a new language and a new style. This coexistence of two distinct approaches corroborates and reinforces both, making them mutually intelligible.

Wölfflin never intended to sacrifice the individual features of a work of art. He knew very well that in art there are the traits of the artist or the geographical and national, which cannot be eliminated.

5 On other occasions, I have conducted comparative studies of Saussure's theory and that of another member of his generation, Bergson. See the article "Alcuni problemi epistemologici relativi al *Corso di linguistica generale*" [Epistemological problems in *Course in general linguistics*], in *Lingua e stile*, 3 (1968).

In some instances, he even evoked the objectionable notion of race,[6] just as Saussure admits that language is actualized or used differently by individual speakers, groups, or communities that are distinguished by professional interests. There is an infinite number of idiolects within each language, but this does not exclude the fact that to understand them, we have to keep the "general" model in mind; otherwise, we sink into a bog of shapeless, unrelated, and incomprehensible events.

Another element that can unite Wölfflin and Saussure is the double, possibly contradictory, feature inherent in their position. They agree (from a distance and without knowing each other) in their acknowledgment of the generalizing impulse. We have to read the general lines beyond the multiple phenomena and to understand how such an impulse cannot have limits. From the style of artistic (or linguistic) phenomena, we shift to that of ethics, customs, and so forth, up to the general lifestyle of an epoch: sectorial styles take shape within a cohesive super-style, finding some of the same relationships that, in each, regulate the encounter/clash between individual phenomena and the collective experience. An analogous tendency applies to Saussure. We know that he endorsed an expanded use of the concept of language, to the point of making it synonymous with any organized system of symbols; so much so that, rightly or wrongly, semiotics, a discipline that generalizes to the maximum degree, emerges from this. At the same time, both "operators" kept to the specific. Wölfflin almost never attempted to establish explicit homologies between the styles he found in art and those in other cultural systems (and in other artistic, scientific disciplines or even at other levels of culture, for example, those of material culture or technology). The same is true for Saussure who hinted at a possible extension of his concept of language, where the entire cultural life of an epoch is considered to be a system of languages inserted into one another (though he never took this step). However, both authors recognized the existence of a powerful impetus to extend the generalizing process and not to stop at the limits of specific individual fields.

Moreover, they never resolved another ambiguity for the good reason that it is not possible or even useful to resolve it since basically such ambiguity is fertile and worth preserving. Are the forces that acti-

6 See Heinrich Wölfflin, *Principles of Art History* (New York: Dover Publications, 1932).

vate this generalizing tendency internal or external? Do the phenomena of a given discipline come together, meld, or become synthesized within the general lines of a style due to internal factors or external constraints? For example, do economic, technological, material factors make their impact detectable within the boundaries of the various areas of production in "high" culture (like those of art) and, therefore, determine the dual processes discussed above: i.e. the coming together of individual phenomena and their changing together, in relation to the style of a preceding epoch? In other words, do styles change because such conditions as weariness and listlessness emerge internally, from which springs the need to look for new avenues capable of stimulating the senses in more emphatic ways? Or is it the case that the changed conditions in material culture (the economy and technology) influence styles and cause them to change? Wölfflin did not address these issues (the same ambiguity is also retained by Saussure, in his own discipline). Both certainly rejected any univocal "explanation" that results in the subordination of one discipline to others that are considered "stronger." They saw in this a consequence of the genetic bias dear to Positivism; that is, the pretension of explaining the complex by starting with the simple and explaining the ideal by starting with the material. In the end, the dignity of general forms, if it had to be considered to be derived from concrete, individual, material givens, would be destroyed. The fact remains that, in order to avoid genetic-causal biases and to maintain a healthy respect for the limits of their respective disciplinary fields, both limited themselves to describing or establishing formal typologies and avoided the dilemma of deciding whether the factors of change are internal or external. It is however the case that, even as an outline, both suggest the possibility of ideal extensions and show how that same generalizing impulse compels us to look for links or find correspondences in the various styles and epochs. Perhaps today we can try to solve the problem, not by privileging the internal factors of change at the expense of the external ones or vice-versa, but by granting to each its rightful space. This is what culturology applied to the phenomenology of styles aims to do.

THE WORKS OF WÖLFFLIN

Turning now to the specific work of Wölfflin, we should recall that, at about the age of 20, he found himself attracted to the field of experimental psychology, which we should see as one of the closing chapters of Positivism, a genetic, analytic, and atomistic approach that is the opposite of what would be the methodology of the German scholar (and his "homologues"). On the other side of the coin, this too is a characteristic of the proponents of general forms perceived *in re*. In fact, they intend to entrust the regulation of the general forms not to an abstract mind but to the physical presence of the body and its sense organs. For Wölfflin, if there is an entity entrusted with the regulation of styles, the point where the styles expressed in different disciplines converge, it is to be located in our body. In a sense, styles in painting and architecture express, by proxy, the way the body of the subjects who invent those styles behaves. This is true on condition that this body in turn is collective, social, and most importantly historical, that is to say, separated from physiology and inscribed in culture – here we shift from physiology to a type of transcendental psychology, which amounts to a phenomenology. This allows us to understand that the first work of the young scholar is, in fact, *Prolegomena zu einer Psychologie der Architektur* [Prolegomena for a psychology of architecture] (1866) where we find the following important statement: "Our bodily organization is the form in which we understand the corporeal forms of art."[7]

Perhaps it is no longer a matter of speaking of experimental psychology, in particular that of Johannes Volkelt. Wölfflin elides this last phase of Positivism to deal with the theory of *Einfuhlung* [empathy] which, following Kant, restores psychology's active and shaping role, one capable of inserting itself in the phenomenal material. This is psychology as the capacity to emanate *a priori* synthetic forms of sensibility (aesthetics). Things or phenomena do not inscribe themselves on a *tabula rasa* [blank slate] or a neutral, compliant receiving device. On the contrary, the affirmation that we see only what we are conditioned to see by history and culture is a constant theme in Wölfflin's work. The eyes (or even the senses of touch and hearing) are not simply physiological apparatuses; they are also, and above all, historical modalities in the process of becoming and they undergo transformations across time. Changing styles correspond directly to the general dispositions of the behaviour of our bodily senses.

7 This is quoted from J. Gannes' useful biographical profile, "Heinrich Wölfflin," in *La critica d'arte* (May, 1950), 1–8.

Just two years later, Wölfflin publishes the book to which his fame is tied, *Renaissance and Baroque*, where he formulates almost fully his theory of bipolar variation. From that point on, his thinking develops through a process of intension rather than extension, which is to say refinement rather than expansion into new areas. *Classical Art: An Introduction to the Italian Renaissance* is published in 1899. (In the meantime, he had succeeded at the University of Basel the famous Jacob Burckhardt, considered one of his teachers, but what a difference between them due to the generation that divides them.) Then, in 1915, we have *Principles of Art History*. For the next thirty years, until his death, he works on perfecting and deepening his ideas, as well as defending his theories from attacks and incomprehension.

In my observations at the start of the present section, I identify two features of styles described by Wölfflin. The first is the generalizing trend, the coalescing of many occasional actions, personal or group choices, within a fairly organic and unified system that allows itself to be fixed and extrapolated. At the same time, history flows and that unity ("language" for Saussure) is replaced by another unity involving a transition that is not arbitrary, but one that allows us to identify its unfolding in recurring typologies. This is how Wölfflin's famous binaries are born, binaries that are fixed at five in the final codification of his principles. (In fact, they are already present in *Renaissance and Baroque*.) As is well known, the binary oppositions are: linear-painterly, plane-recession, closed-open, multiplicity-unity, clarity-obscurity. The examples, explanations, and descriptions with which Wölfflin presents these dichotomies and makes them work in his three most important essays are so abundant that it would be pointless for me to retrace his steps or cite lengthy passages of his analysis. Instead, I offer a series of reflections and comments.

First and foremost, Wölfflin is quick to discard the misguided notion of a history or natural evolution of styles, as though it were progress for the open, unified, and the painterly to be considered nearer to nature than they were in an earlier phase of development. Instead, we are dealing with cultural systems, each with its own *raison d'être* and its own dignity. It is possible to be great artists in one system or the other. This implies full acceptance of historical as well as cultural relativism and the transcending of prejudices toward both primitive and Baroque art, which was viewed as a degeneration and corruption of taste in those years straddling the seventeenth century, like a fatal decline following the heights attained by Renaissance Classicism.

THE SPIRAL GRAPH

We should not draw from this the conclusion that styles succeed one another according to interchangeable and arbitrary criteria. There is what we could call a historical vector or directional flow marked by a transition from closed to open and not vice-versa. This does not mean an improvement; rather it indicates an intrinsic logic of development. Consequently, each of the five binaries indicates a precise developmental direction. In addition, we do not have a law of variation or a "general" form if the succession does not recur in the course of history. Wölfflin found himself faced with the problem of guaranteeing the uniqueness (though it is the uniqueness of "general" forms) of both the Renaissance and the Baroque while at the same time redeploying his favourite device, the variation schema, so that he could apply it to other historical periods. Consequently, he introduced a cyclical feature into the history of styles in order to account for the unfavourable presence of revivals. For example, after the domination by the Baroque and Rococo in the 1600s and the first half of the 1700s, with the resulting triumph of the second term of the binary (the painterly, the profound, the open, etc.),[8] there was a need to wipe the slate clean and to start over again between the end of the 1700s and the first decades of the 1800s. David and Neoclassicism in general signal a return to forms that are closed, rhythmical, articulated, clear, etc. Such a return, however, does not signify the rediscovery of the distant positions of the Renaissance. Wölfflin was too steeped in history to imagine the oscillations of a pendulum that are too repetitive and too symmetrical. As a result, he proposes a more appropriate graph, one that ensures a proper resultant vector for the axis of linear time and for the circular, oscillating motion, namely, a spiral diagram in which the returns occur at different heights and therefore one never crosses the same point.[9] Had he ventured further in his analysis of the art of his times, Wölfflin would have been able to determine that an entire oscillation, from closure to openness, from clarity to obscurity, etc., had already taken place before his eyes (from David, Ingres, and Friedrich to the Impressionists), and that a successive oscillation was

8 Heinrich Wölfflin, *Principles of Art History* (New York: Dover Publications, 1932).
9 Ibid.

already under way or was about to complete its course. Following the Impressionist intoxication (where everything sinks into an orgy of fleeting sensations), the art of the first part of the twentieth century revives the taste for essences, for solid, intact forms (just as, at the theoretical level, Wölfflin himself expresses the need to recover the "general," unified forms that had emerged corroded and threadbare from their immersion in positivist psychologism), except that the clear, closed, etc., forms of Cubism and similar styles are radically different from those found at the start of every cycle (Renaissance, Neoclassicism), apart from this common morphological marker. Thus, the validity of the spiral graph is confirmed. The return is never literal since only certain vectorial elements, certain general morphological features, return. These, however, are to be defined in each cycle with more specific terms tied to given historical contexts. Finally (but this would have been to ask an elderly critic who had dedicated his whole life to historical studies to do the impossible), Wölfflin could have determined that a phase of open forms (painterly, profound, obscure, etc.) was about to get under way in the early 1940s. These forms correspond to the cluster of styles known as Informal, Tachisme, Abstract Expressionism, and such.[10]

But, as noted above, the Swiss scholar placed precise historical and theoretical limits on himself and he was careful not to go beyond these limits by drifting into other territorial waters. From the historical point of view, he considered himself a specialist of modern art, understood in the academic or traditional sense (from about the mid-1400s to the 1700s), and he expressed his complete reluctance to surpass those chronological boundaries (leaving aside the aforementioned references to the clean slate followed by the emergence of a "contemporary" cycle stretching from Neoclassicism to Impressionism). Even more noteworthy are the self-imposed theoretical limits. Wölfflin never undertakes an interdisciplinary inquiry; therefore, he does not ask if the styles of the visual arts establish homologies in other more or less contiguous disciplines. Nor does he ask if or how, through a series of mediations, they come to interfere with the stratum of material culture. However, each time I came up against these

10 In this regard, I would like to point out that I entered the world of art criticism during the time of the Informal and I started out by applying the binary closed-open forms to that style. See, "La XXVII Biennale" [The 27th Biennial Exhibit] in *Il Verri* (1956), no. 1.

limits, which are consciously accepted by the author, I was quick to observe that he never failed to indicate the ways in which these ideas could be extended.[11] If nothing else, I have argued that, according to his theory, a synthesis of the various styles of an epoch is located in a kind of corporeality and transcendental behaviourism of a collective subject (a social body). It appears to me entirely legitimate for those of us who follow Wölfflin's teachings, to continue along the path he blazed and to attempt to carry out these expansions and go beyond both historical limits – if for no other reason than the fact that the history of styles has proceeded and new material on which to shine the interpretive light has accumulated – and interdisciplinary limits since we have greater awareness and more precise instruments at our disposal. As for the possible accusation of being presumptuous and "biting off more than we can chew," we could defend ourselves by citing the adage of the dwarves who stand on the shoulders of giants and only because of this are taller by a span, or we could cite the equally proverbial: "Poor is the pupil who does not surpass his teacher." In other words, we have to take into account the effects of inertia whereby an object moves when it is pushed.

Wölfflin's historical and theoretical limits certainly contain lessons that can be used to project forward. It is true that he essentially confined his inquiry to two or three centuries of the history of western art, but within that narrow segment, he was able to demonstrate his great ingenuity on two occasions and to apply his vectors, with a degree of superimposition. In fact, blending the results of his historical works, *Renaissance and Baroque* and *Classical Art*, yields the following correlation: the art of the second half of the 1400s is to the 1500s as the art of the 1500s is to the 1600s. In other words, if we compare Leonardo, Raphael, Andrea del Sarto, Fra Bartolomeo, Titian, and the northern artist Dürer, to Botticelli, Lorenzo di Credi, Ghirlandaio, and Verrocchio, they are relatively "open," free, and painterly. Hence, the second term of Wölfflin's binaries applies to them, whereas if we compare them to the Carracci family, Rubens, Bernini, and Rembrandt, the first term of the binary applies since their works appear to be relatively closed, linear, symmetrical, rigid, clear, analytical, and so forth. In sum, both terms of each binary are to be understood as rela-

11 In this matter, Wölfflin could have gone all the way back to the founder of all phenomenologies of style, Vasari, who, in the famous *Proems* to his *Lives*, identifies three waves or styles in the evolution of the Renaissance.

tional and not absolute. The terms derive their meaning from the position or the vector to which they belong. This fact serves to blunt the usual accusations from nominalists and idealists (two equal and opposing groups that are in practice united in their effort to invalidate Wölfflin's typologies with their excessive attachment to individual historical entities). They would argue that a successful work of art is always by definition unified and closed, but at the same time it must possess an appropriate amount of multiplicity in order to avoid becoming monotonous. In fact, each time the notion of "open" was applied in historiography, the objection was raised that it contradicts the qualitative nature of the work to which the notion is applied. This kind of thinking leads to the ineffability of the work or the tautological statement that the work is what it is. It is no accident that a loyal follower of Croce, Carlo Antoni, objected to Wölfflin, saying "omnis determinatio est negatio" [every determination is a negation].[12] Wölfflin could have replied that only partial negations, differences, divergences, and comparisons allow us to understand the various styles. The cult of "everything in everything" leads to the melting pot concept where everything blends together. At best, it would be possible only to make "deictic" gestures that indicate either the presence or the absence of poetry or art. (All in all, Croce's method can be reduced to this.) According to Wölfflin, "open" and "closed" forms are, instead, situational or relative qualifiers whereby, a more open level can follow an open level, causing the latter to slip into the "closed" category. Raphael is "open" relative to Botticelli but "closed" relative to Bernini or an artist like Annibale Carracci, even if he claims to emulate Raphael. While on the diachronic axis it is worth emphasizing differences, at least if we want to understand and interpret, on the synchronic axis it is appropriate to minimize the differences between individuals, groups, schools, countries, or at least not to use these differences to prevent the identification of a general stylistic template that does not erase those distinctions. In essence, this is the dialectic that Saussure institutes between acts of *parole* and the presence of *langue*. In an epoch that is relatively closed it is possible to be closed in a thousand ways, with some advance and some lag, in relation to the median or the resultant vector for the epoch. With his or her dis-

12 Carlo Antoni, *Dallo storicismo alla sociologia* [From history to sociology] (Florence: Sansoni, 1940), 229.

cretion, the historian determines whether an emerging stylistic feature still fits the template or "diverges" too much and so begins to constitute a new style. Every device allows a certain play of its constituent parts that does not compromise the function of the device, but beyond a certain limit, if the play is excessive, the mechanism breaks down. Raphael's work is graceful, fluid, and therefore "open" although it is so within the "closed" nature of typical works from the first part of the 1500s. Correggio, however, is even more "open" to the point where he can no longer be reintegrated. In fact, Wölfflin sees him as a transitional figure that moves toward the next set of "open" forms, i.e., the Baroque.

Naturally, if we want to stay within the confines of strictly historiographic thinking, we need to acknowledge immediately the limits that Wölfflin could not avoid due to the nature of the studies of his day. These limits, however, do not compromise the overall validity of his methodology. As indicated by the many comments scattered throughout his writings, he was largely unaware of the existence of the mannerist phenomenon, which was re-evaluated in the years following his own development as a scholar. (In this regard, we could mention his relative imperviousness to subsequent developments, as though he stopped at the end of the 1800s, remaining unresponsive to later achievements, since he was concerned with developing his own thinking.) Today, we know that the styles of the 1500s and 1600s were much more fluid and complex than he realized. In a sense, he almost lost the opportunity to expand the applicability of his vectorial schema. In fact, he was not aware of the new "closed" phase that is Mannerism (Parmigianino, Rosso, Pontormo, etc.), which developed in the third and fourth decades, then triumphed in mid-sixteenth century. He was likewise unaware of the "return to closed forms" of the neo-renaissance phenomenon of the last decades of the century when, especially in connection with the spirit of the Counter-Reformation, there was a desire to revive simple, archaic, more static, and more humble forms that responded to the need for verisimilitude and conformity to nature. Essentially, the inadequate historiographic framework available to Wölfflin did not permit him to "animate" with peaks and valleys, as well as polar swings, the development of an entire century and led him to imagine the whole century as an uninterrupted flow or one long transitional phase into the 1600s. Even with respect to the 1600s, he undoubtedly had a much more schematic and unified vision than we have today since we have seen the intensification of the clash of several currents: Caravaggio's naturalism, the "reform" and classical landscape of the Carracci family, the work of Poussin, and the Baroque, understood in the more restricted sense as stretching from Bernini to Pietro da Cortona. Even here, is it better to flatten the century into a single image or to enliven it with crests and troughs? After the excessive openness of the Baroque, in the narrower sense, do we not have a return to order in Poussin and his peers? (In this case, the phenomenon of the Carracci family would be synchronic or rather prior to the twisted, dramatic forms of the Bernini school.)

Paradoxically, the lack of articulation in Wölfflin's historical approach, owing to the fact that studies were not fully developed at the time, allows him to "jump" all the way to a recent and more advanced and sophisticated level. Whereas in the past, the various definitions of the Baroque, each more or less limited in scope, vied with each other to be considered the only valid one or resisted being excluded by other definitions, what prevails today is a pacifying approach aimed at conciliation that allow Naturalism and Classicism to coexist under one umbrella. This very broad and inclusive notion of the Baroque may also be considered another victory by the generalizing impulse – a view that Wölfflin always espouses. The fact remains that the five binaries, within a coherent system, "cover" the various characteristics of each sub-tendency of the Baroque age. In the end, we are inclined to see the particularities of each school as instances of *parole*, or circumscribed styles, so as not to obstruct the identification of a common physiognomy or a set of features that are sufficiently recurring and verifiable throughout the Baroque age.

"INTERNAL" AND "EXTERNAL" FACTORS

In addition to historiographic caution, Wölfflin has always demonstrated similar caution as regards straying into other disciplines, to the point of creating the impression that, for him, stylistic variance has a completely internal genesis. In truth, he always linked the presence of internal factors of change to external factors.[13] That is, he never really

13 This was also the impression of Erwin Panofsky who, in his essay *The Problem of Style in the Visual Arts* (1915), is critical of Wölfflin's excessive emphasis

denied that the stylistic vicissitudes of the arts interact with those of other fields, to the point of being part of the wider, general style of the body and its sense organs. What these other disciplines are and how the interaction comes about are questions he never wanted to consider. It is clear that to establish a homology between styles in the visual arts and styles in material culture, I have been applying the theories of McLuhan, according to which, in the 1400s, there is a substantial homology between Gutenberg's invention of the movable type printing press and the discovery of Renaissance perspective, which is largely the work of Alberti.[14] That moment signals the beginning of the "modern" period, whose unity and continuity derive from the parallel development of mechanical industrialism (the machine that can produce "copies") and the graphic representation of space governed by the principle of the *camera obscura*, which was consolidated by the advent of photography. However, within the broad lines of this general uniformity, there are many subsequent phases, stages of adaptation, moments of nostalgia for the past, as well as moments of rapid advancement. We know that for McLuhan the printing press was a typically "hot" medium because it leads to the hypertrophic development of optical-perceptive values at the expense of other values. In short, from his theories we can derive a binary or a vector continuum that extends from "cold" (the predominance of tactile, acoustic, bodily values in general) to "hot" (distancing of the object and control of the object through sight, with intellectual gain but affective loss).[15] We also know that, at the end of a complete cycle, for McLuhan too

---

on the internal or formal aspects of style at the expense of the external aspects relating to the other disciplines of an epoch. The accusation of formalism also comes from Croce but for different reasons. For Panofsky, in addition to observations pertaining to the internal aspects, it is important to consider the external ones (in which he himself specialized, having developed the notion of iconology). For Croce, neither aspect is relevant because the synthesis of form and content is an ineffable unity. In order to understand Wölfflin, we need to keep in mind his broadly Kantian approach whereby form is always for content. It is an *a priori* of sensory perception. Thus, by emphasizing the form, Wölfflin cannot refer to the material content that the form is capable of selecting.

14 See Chapter 1.

15 This methodology of the binaries is discussed in detail in Luciano Anceschi's phenomenology, beginning with his famous essay *Autonomia e eteronomia dell'arte* [Autonomy and heteronomy of art] (Florence: G.C. Sansoni, 1936).

there is a wiping the slate clean and a starting over again, at least if developments in material culture (in technology) permit it. The contemporary age, which is based on the massive deployment of electrotechnics, witnesses a "return" to tactile values or to a tactile-organic relationship with the things that go against the visual space of Renaissance perspective. However, McLuhan clearly recognizes that the coincidental institution of the printing press and perspective did not make its effects known overnight. For quite some time, the dominant culture remained tactile-oral (acoustic space, mouth-ears circuit). Wölfflin had already come to a similar conclusion in his studies on the styles of the second half of the 1400s in Italy, which are supposedly hard, analytical, and excessively detailed, expressing the tactile attention of the artist to each particular of the anatomy and the drapery.[16] There is no atmosphere, distance, or depth in the paintings of Botticelli or Ghirlandaio because everything is experienced at close range. The viewer does not see the object in a graduated scale. Essentially, what we have is an "approach" that is primitive, archaic, "cool" in McLuhan's sense. Then, we have the great revolution of Leonardo, which consists of exploiting distance and rejecting the tactility of the body, relying mainly on perceptual mediation and generally equating fullness and emptiness, the body and the air. This is so even if the first part of the 1500s is only a stage. Commensurate with the much more advanced achievement of the 1600s (the Baroque, in Wölfflin's broader sense), this phase exhibits characteristics that are still relatively "closed," that is to say related to the plasticity of bodies that are not yet completely depicted in depth and are not yet totally embedded in the background. To conclude, modern humanity comes into existence gradually and if we suddenly find the use of certain implements typical of modernity (the printing press at the material level and per-

16 *Arte classica* [Classic art] Ital. trans. (Florence: Sansoni, 1978), for example, 207, but also the ample and informative analyses that follow. We can sometimes note a bias in Vasari as regards Wölfflin's nationality. For the proto-historical Italian, the Renaissance is a unity despite the fact that it unfolds in three *maniere* (see note 12). For the representative of German culture, there is a caesura between the "early" and the "late" Raphael. Let us recall that a century earlier there was the Nazarene phenomenon (followed by Pre-Raphaelitism), which sought to revive the "closed" forms of the 1400s. For German and Anglo-Saxon historiography (we need only think of Ruskin), the corruption associated with "open" forms begins with Raphael.

spective at the theoretical level), many decades or centuries have to pass before they imprint themselves on the collective sensibility and succeed in shaping it thoroughly. However, from this comparative analysis, it is evident that Wölfflin's typologies are congruent with McLuhan's, even though each functions within its own sphere (but there is also the possibility of exchanges between the two).

Just as it is possible to establish congruence or homology between the internal stylistic field of art and the external field of material culture, so it is with the intermediate fields of the other sciences.[17] In fact, in philosophy and literature, "modernity" does not come about all at once. In the 1400s and the first part of the 1500s, what prevails is a humanistic concept of humankind as a protagonist who occupies a privileged position in the cosmos and acts boldly by virtue of the powers of eloquence. Both scenarios explain the reason for which fifteenth-century painting is marked by egocentrism or anthropocentrism. The human being rules supreme and strikes poses of studied eloquence. Here too, through the technique of *sfumato*, Leonardo's revolution introduces a tentative notion of cosmocentrism, which leads humans to tone down their privileges. Once more, the fusion of human presence with the cosmos that we see in the first part of the 1500s is relatively insignificant in relation to the dark, painterly, open, profound characteristics of works from the 1600s. As we can see, factors that are internal only to the visual arts field can interact with external factors, both at the level of technology and at corresponding levels of the other disciplines, the keepers of the knowledge of an epoch. Though he refuses to sort out these complicated relationships, Wölfflin still allows the principal structures of his interpretive system to emerge, a foundation on which later successful developments would rely.

17 This is the kind of integration in which Panofsky excels, especially in his classic study on perspective.

Postscript

When this work first appeared in Italian, the threshold of the third millennium was still some distance away and when the second edition came out, we had just crossed into it. Now that we are more than a decade into the new century, I feel the need to provide an overview of where things stand. In general, we could say that the revolutionary ideas of 1968 have spread like wildfire. As we have seen in the preceding pages, there was a revolution in that year, one that affected technology primarily. We became aware of the fact that technology, particularly electronics, was becoming dominant and was starting to elude the control of the western world. As usual, the most interesting phrase comes from McLuhan: his famous "global village." The networks of electronic connections, in other words the Internet, broke down the borders between states, between continents, connecting the entire globe through an instantaneous communications system.

McLuhan's metaphor of the global village is especially useful because it contains two opposing meanings that constitute an oxymoron of sorts. There is the concept of globalization with the attendant risk that the process could lead to homogenization: the whole planet eating the same foods, enjoying the same entertainment and pastimes, and acquiring the same gadgets. However, the concept of the village introduces quite a different notion, almost confirming the adage "small is beautiful" by referring to a relatively small physical space in which the members of a community live in close contact and dialogue with one another. Therefore, this process does not lead to the elimination of local characteristics; rather, it accentuates and ensures their survival. As a result, we are able to merge the two semantic spheres by using the word "localism" to suggest this possible marriage of opposites. I have already mentioned that the revolution of 1968, particularly in the visual arts, produced the so-called "death of art," which is to be taken literally as the end of a certain procedure that Western culture in the modern age had imposed on the entire world, that being the notion of the painting as a piece of canvas fastened to a frame that is rigorously rectangular, corresponding perfectly to the Gutenberg Galaxy.

I have also emphasized the fact that, although they broke with some conceptions of this kind, the historical avant-gardes did not succeed in coming completely out of the medium of painting. This was also true for such innovators as Paul Cézanne, Pablo Picasso, Umberto Boccioni, and Piet Mondrian. It would take a long time for the technological revolution to permeate the sensibility of people in general and to become completely accepted. We know very well that paintings executed on canvas and intended to function like windows onto external reality were incompatible with other cultures not affected by the printing press. These cultures had to adapt to this development or be left out of the process and continue to use obsolete forms of expression. After 1968, the West provided the example to follow. The painting was either avoided altogether or executed on open and unlimited media, assuming modes of expression described by such terms as graffiti art and mural art. Also broken is the old western taboo against the use of letters of the alphabet or writing in general in the visual arts. Finally, even three-dimensional works were entrusted to so-called installations, created by using materials taken directly from the physical world and assembled in a variety of ways. This includes paying special attention to the spaces designed to house what are now referred to as "site specific" installations. While they seemed to be totally out of place in the good traditions of the West, from the Renaissance to the modern period, such practices were extraordinarily well suited to non-western cultures but were renounced or abandoned when the western model of painting, understood as faithful reproduction of reality, came to dominate.

Therefore, the adoption of these extra-artistic procedures, which were alien to what had been done in the West for many centuries, rather than flatten artistic production in the world or reduce it to homogeneity, prompted various countries and cultures to re-appropriate their own history, thereby offering a range of important artistic products as well as significant levels of creativity. Clearly, countries that entered most quickly into the technotronic age, for example,

Japan, South Korea, and China, were the first to produce important results in the artistic field as well. Even the macro-continent of India and the countries of Latin America, following the lead of the giant Brazil, have begun to respond to such a powerful stimulus. And we should not forget the Near East, South Africa, and Australia. In addition, technological innovation has facilitated access to this creativity with respect to the other half of the human race, namely women. It is no longer possible to list all the cases of women artists who have come to the fore, for example, Iranian Shirin Neshat, Japanese Mariko Mori and Tabaimo, and Lebanese Mona Hatoum, even if these impressive careers benefitted almost always from time spent in important western cities, such as London, New York, and Los Angeles. The twenty-first century promises to develop widespread magnificent creativity to which all the countries of the world may contribute by surpassing every limitation or exclusion on the basis of sex, religion, or ethnic origin.

Annotated Bibliography

CHAPTER ONE

Henri-Jean Martin's *Histoire et pouvoirs de l'écrit* [The history and power of writing] (Paris: Librairie Académique Perrin, 1968) is an extensive study of the material aspects of writing (its forms and structures throughout the various historical and technological periods) with particular attention paid to the notions of *ductus* and handwriting. In terms of an approach to the concept of culture that privileges its material stratum, *Culture: A Critical Review of Concepts and Definitions* (New York: Vintage Books, 1969) by Clyde Kluckhohn and Alfred Louis Kroeber is fundamental. Also useful is Marvin Harris' *The Rise of Anthropological Theory*, the sub-title of which, *A History of Theories of Culture* (New York: Crowell, 1968), is especially informative. I have followed Harris' thoughts closely in the present volume. He is not only an impartial historian of cultural anthropology, but also a lucid and passionate proponent of the primacy of "cultural materialism," which is also the title of his most important essay, as far as methodology is concerned, *Cultural Materialism: The Struggle for a Science of Culture* (New York: Vintage Books, 1980). As for Harris' many and always fascinating applications to various historical-cultural periods, we need only consult *Cannibals and Kings: Origins of Cultures* (New York: Vintage Books, 1978). More recently, this American scholar narrowed the range of his critical attention somewhat as he examined our own age in *America Now: The Anthropology of a Changing Culture* (New York: Simon and Schuster, 1982). The application of his approach has become less convincing, thereby revealing the difference between cultural anthropology and culturology or the science of culture. The former is appropriate for discussions on remote cultures, for instance, prehistoric communities, whereas the latter has instruments "calibrated" in such a way as to enable it to intervene in epochs closer to us, such as the advent of industrial society as well as the crisis associated with the transition to the post-industrial era (I return to this point in the pages that follow). As for the science of culture or culturology in the strict sense, a pivotal work is *The Science of Culture: A Study of Man and Civilization* (New York: Farrar, Straus and Cudahy, 1949) by Leslie White, an author I follow particularly closely because of the subtle methodology he employs in describing the interconnections between the stratum of material culture and the stratum of "high" culture, which he relates especially to symbolism. Another family of culturologists to which the present study is greatly indebted, is the one established by Harold Adams Innis with *The Bias of Communications* (Toronto: University of Toronto Press, 1951), where the notion of communication itself is understood in a very "strong" and fertile way encompassing as it does all possible involvement of the media, both material and symbolic. The work of Innis, who is not well known in Italy and whose work was translated with some delay with respect to the publication of the original, was expanded very successfully by McLuhan, whose complete writings are accessible to the Italian reading public. These include *The Mechanical Bride: Folklore of Industrial Man* (New York: Vanguard Press, 1951), a work written early in his career. At this point, the Canadian scholar's theoretical horizon is still limited to industrial-mechanical culture and its most conspicuous product, the automobile. McLuhan subsequently places the same mechanical era into historical perspective by relating it to the "machine" invented much earlier, the movable type printing press, in the first of his two masterpieces, *The Gutenberg Galaxy: The Making of Typographic Man* (Toronto: University of Toronto Press, 1962), translated into Italian a decade later. His other masterpiece was translated into Italian as *Gli strumenti del comunicare* [The instruments of communication] in 1967, a few years after the publication of the original. The distortion in the translated title is regrettable since the original title, *Understanding Media: The Extensions of Man* (New York: McGraw-Hill, 1964), places the emphasis on the concept of media, which is fundamental in terms of a culturological perspective – among other things, for the ambiguity of the term, which can refer both to material and symbolic or "high" culture media. (The current concept of "communication," instead, risks privileging the latter over the former.)

Finally, Eric McLuhan, the author's son, published *Laws of Media: The New Science* (Toronto: University of Toronto Press, 1988) following his father's death in 1980; this is an ambitious attempt to provide a definitive synthesis of the far-reaching nature of Marshall McLuhan's theories, using notes Eric's father had left him. Also useful is the monumental correspondence edited by Marshall's wife Corinne and Marie Molinaro, a selection of which has also appeared in Italian translation with the assistance of F. Valente (1991). This correspondence confirms the close friendship between McLuhan and Walter Ong (deceased in 2003), the author of seminal studies on the impact of the electronic media on the entire system of forms in "high" culture. See especially *The Presence of the Word: Some Prolegomena for Cultural and Religious History* (New Haven: Yale University Press, 1968), *Orality and Literacy: The Technologizing of the Word* (London and New York: Routledge, 1982), and *Interfaces of the Word: Studies in the Evolution of Consciousness and Culture* (Ithaca, New York: Cornell University Press, 1977). The first and third works include an introduction by Ong himself.

Moving from the weighty contributions of North American scholars to those of European thinkers, I feel obliged to acknowledge the great debt owed to Lucien Goldmann and his notion of homology, which ensures a continuous exchange on the basis of "equal dignity" between the two strata of culture, the material and the symbolic, thereby putting an end to the lengthy history of subordination of one to the other, which saw the symbolic sphere either sacrificed in favour of the material (as occurred during the long tradition of Marxist studies, from which Goldmann himself emerged) or reduced to "being embarrassed" by its inconvenient and crude partner, to the point of removing or concealing it. Goldmann's best-known work is *Pour une sociologie du roman* [Towards a sociology of the novel] (Paris: Gallimard, 1964), but also noteworthy from the standpoint of theory are his earlier contributions, *Introduction à la philosophie de Kant* [Introduction to the philosophy of Kant] (Paris: Gallimard, 1948) and *Recherches dialectiques* [Dialectical investigations] (Paris: Gallimard, 1959). Although he rightly placed the material stratum on the same level as the ideal-symbolic stratum, Goldmann erred in identifying the material stratum only with the industrial means of production, excluding the broader and more fluid concept of media provided by North American thinkers, McLuhan in particular. The reason for this oversight may be the fact that McLuhan lived closer to and just before the transition from the "mechanical bride" to the electronic era.

In France, the so-called Annales School of historiography had a large following. Its leader was Fernand Braudel (e.g., *Capitalism and Material Life*, 1962), an author with a vast intellectual horizon, but with a less reliable and coherent methodology than we find in other schools, which I pointed out in my review (*Il Mulino*, May–June, 1982) of *Nella trama della storia* [In the web of history] (1990), a review that raised the ire of the author of the book, Renzo Zorzi, a staunch defender of Braudel.

Also relevant here is the vast and ongoing debate on the concept of the postmodern, two aspects of which relate to the present study. The first is one of form since I prefer to use the term postmodern in the place of the more ambiguous "contemporary," while the second is one of substance since one of the most important tasks of culturology is to define the postmodern era and with it the material-technological stratum, which should make postmodernism essentially the triumph of electrotechnics and its successors. For an introduction, with bibliography, to this topic, I refer to my book, *Il ciclo del postmoderno* [The postmodern cycle, 1987]. See also my article, "La tecnologia nel rapporto tra moderno e postmoderno" [Technology in the relationship between the modern and the postmodern] in *Forme e pensiero del moderno* [Forms and ideas of the modern], edited by Franco Rella (1989). For the concept of the post-industrial period, which is almost synonymous with the postmodern period, see Daniel Bell's *The Coming of Post-industrial Society* (New York: Basic Books, 1973). Equally extensive is Zbigniew Brzezinski's *Between Two Ages: America's Role in the Technetronic Era* (New York: The Viking Press, 1970). Also very interesting and successful are the "scenarios" proposed by Alvin Toffler in *The Third Wave* (New York: Bantam Books, 1980). An indispensable point of reference, but one whose objectives are not always clearly articulated, is Jean-François Lyotard's *La Condition postmoderne* [The postmodern condition] (Paris: Éditions du Minuit, 1979). In addition to these broad culturological studies, there are others that are more specific and deal with the various fields in which the concept has emerged. See, for example, American critic, Ihab Hassan's *The Dismemberment of Orpheus: Toward a Postmodern Literature* (Oxford: Oxford University Press, 1971). The concept is extended to architecture by Charles Jencks in *The Language of Post-Modern Architecture* (New York: Rizzoli, 1977). The rich debate on this subject in North America is well synthesized by Linda Hutcheon in *A Poetics of Postmodernism* (London and New York: Routledge, 1988). See also a recent contribution of mine titled, "Dal moderno al postmoderno" [From the modern to the postmodern] in *Storia generale della letteratura italiana* [General history of Italian literature] (Milan: Motta, 2004, vol. xii).

CHAPTER TWO

The foundation for this line of thinking is my reliance on the "philosophy of symbolic forms" formulated by Ernst Cassirer in the homonymous work published between 1923 and 1929. Of critical importance is Susanne Katherina Langer's application of this approach to the problem of the arts and their relationships with the symbolic systems used in the sciences, *Feeling and Form: A Theory of Art* (New York: Scribner's, 1953). On this complex problem see also my own *Corso di estetica* [Course on Aesthetics] (Bologna: Il Mulino, 1989 and English translation, *A Course on Aesthetics*, Minneapolis: University of Minnesota Press, 1993).

As for the formal sciences of discourse, the source is Aristotle's *Organon* since the different histories of analytics and dialectics stem from this treatise. For analytics, see Robert Blanché, *La logique et son histoire d'Aristote à Russell* [The history of logic from Aristotle to Russell] (Paris: Armand Colin, 1970); *Aspetti storici, filosofici e matematici della logica moderna e delle sue applicazioni* [Historical, philosophical, and mathematical aspects of modern logic and its applications], edited by E. Agazzi and C. Cellucci (Rome: Istituto dell'Enciclopedia Italiana, 1981); and Heinrich Scholz, *Concise History of Logic* (New York: Philosophical Library, 1961). The hostility and prejudice we frequently find today directed at the humanistic sciences in the name of the primacy of analytical reason are described incisively by C.P. Snow in *The Two Cultures* (Cambridge: Cambridge University Press, 1960). A balanced defence of dialectics and its inevitability is conducted by Georges Gurvitch *Dialectique et sociologie* [Dialectics and sociology] (Paris: Flammarion, 1962). I should also cite Goldmann's aforementioned *Recherches dialectiques*. In this regard, a fine theoretical contribution that argues the inevitability of these two logical pathways is Enzo Melandri's *La linea e il circolo* [The line and the circle] (Bologna: Il Mulino, 1968). In a similar vein, the role of rhetoric is of fundamental importance in ascribing to the human sciences the dignity of logic but without fixing them within the rigid mold of analytics. Aristotle excludes the "science of discourse" from the *Organon*, but he devotes a separate treatise to the subject. From here, rhetoric spreads in western culture across the centuries, though with alternating phases of popularity and unpopularity. On this topic, see my historical and theoretical comment in *Retorica* (Milan: ISEDI, 1979, now Bologna: Fausto Lupetti, 2011 and English translation, *Rhetoric*, Minneapolis: University of Minnesota Press, 1989). In modern times, rhetoric reached its lowest point until it was rediscovered by, among others, Chaim Perelman in *Traité de l'argumentation: La nouvelle rhétorique* [The new rhetoric: A treatise on argumentation] (Paris: Presses Universitaires de France, 1958), written in collaboration with Lucie Olbrechts-Tyteca. A similar argument aimed at giving deserved space to both analytical reason and rhetoric is made effectively in Italy by Giulio Preti in *Retorica e logica* [Rhetoric and logic] (Turin: Einaudi, 1968), and in the United States as well, for example, by Ivor Armstrong Richards in *The Philosophy of Rhetoric* (London: Oxford University Press, 1936). The intersection of this new sensibility in eloquence and oral communication, on one side, and electronic technology, on the other, is at the heart of the works by Walter Ong mentioned earlier.

The problem of long-distance collaboration between "operators" who belong to different disciplines and cultural contexts geographically (but not temporally) distant from one another is present in the works of Leslie White, also mentioned in chapter 1. I have provided numerous applications of his theories, especially as regards fictional narrators of the first part of the 1900s and their relationship with the great epistemological innovators of the same period. See my article, "Il comico in Bergson, Freud e Pirandello" [The comical in Bergson, Freud, and Pirandello] in *L'enigma Pirandello* [The enigma of Pirandello] (Ottawa: Biblioteca di Quaderni d'Italianistica, 1988). Prior to that, I had published *Comicità in Kafka* [The comical in Kafka] (Milan: Bompiani, 1981) using the same interpretive key.

CHAPTER THREE

The importance of the search for the new is underscored sufficiently well by all the exponents of the best contemporary aesthetics. See in particular John Dewey's *Art as Experience* (New York: Minton and Balch, 1934), which is faithful to the author's methodology and, as such, it recognizes that all human activities are rooted in biology and culture – matrices that can be found in the general concept of experience. In the European context, see as well the best representative of Prague School functionalism, Jan Mukarovsky, *Aesthetic Function, Norm, and Value as Social Facts* (Ann Arbor: University of Michigan Press, 1979) and by the same author, *Studie z Estetiky* [Studies in aesthetics] (Prague: Odeon, 1966). On the subject of the importance of the "new" in aesthetics, see also my work, *Per un'estetica mondana* [Toward an earthly aesthetics] (Bologna: Il Mulino, 1964), as well as the subsequent expansion of these ideas in *Corso di estetica* [A course on aesthetics], previously cited.

For the Freudian concept of the generational clash between fathers and sons, rather than refer to each of the works of the founder of psychoanalysis, I recommend Jean Laplanche and Jean-Bertrand Pontalis's *The Language of Psycho-analysis,* trans. D. Nicholson-Smith (New York: Norton, 1974), and specifically the entries for "Castration Complex" and "Oedipus Complex." The notion of the possibility of symbolically killing the father without replacing him, in an effort to establish a culture free of authority figures, is supported by "leftist" Freudian Alexander Mitscherlich in *Society without the Father* (London, Sydney: Tavistock Publications, 1969). For a systematic application of the generational concept to the history of art, I refer the reader to the following works of mine: "Il Simbolismo" [Symbolism] in *L'arte moderna* [Modern art] (Milan: Fratelli Fabbri, 1966, 2 vols.); *L'arte contemporanea* [Contemporary art] (Milan: Feltrinelli, 1984 and 2005); and "Le generazioni del Quattrocento" in *Studi in onore di Giulio Carlo Argan* [The Generations of the 1400s in Studies in honour of Giulio Carlo Argan] (Florence: La Nuova Italia, 1994); in addition to the applications provided in chapters 4 and 5 of the present volume. I also frequently applied such a concept in my last essay, summarizing all my experience as a teacher, *Arte e cultura materiale in Occidente. Dall'arcaismo greco alle avanguardie storiche* (Turin: Bollati Boringhieri, 2011). In literature, consistent use of this paradigm is made by Romano Luperini in *Il Novecento* [The 1900s] (Turin: Loescher, 1981, 2 vols.). Concerning the question of patterns of stylistic variation and their historical origins, which is largely the focus of the schools of "pure visibility," which are active toward the end of the last century, a good starting point is provided by Lionello Venturi's *Storia della critica d'arte* [History of art criticism] (Turin: Einaudi, 1964, ch. x, "La critica d'arte e la pura visibilità" [Art criticism and pure visibility] with sections dedicated to Hans von Marées, Konrad Fiedler, Adolf von Hildebrand, and Alois Riegl). Venturi himself recognizes the fact that the most outstanding representative of that climate is Heinrich Wölfflin, to whom he dedicates the most ample section of his study. The latter is also the subject of my own reflections, such that I have dedicated a chapter to his work. Of great importance is the School of Dynamic Formalism established in France by Henri Focillon, whose *Vie des formes* [The life of forms] (Paris: Presses Universitaires de France, 1934) remains irreplaceable. George Kubler continues this work in *The Shape of Time: Remarks on the History of Things* (New Haven: Yale University Press, 1962). The use of bipolar schemas is also present in the essays of McLuhan, in the form of the "hot-cool" binary opposition. See more or less all his major works, cited in chapter 1 of the present study. In addition, Umberto Eco's *L'opera aperta* [The open work] (Milan: Bompiani, 1962) can be considered another product of this binary methodology, which, in this case, emerges from a purely formal environment to look for connections with other sectors of the sciences without, however, resorting to the notion of "homology." He prefers to employ the less constraining expression "epistemological metaphor." For a general methodology that takes into account the inevitability of stylistic variation, we have the classic works of Luciano Anceschi, such as *Le istituzioni della poesia* [The institutions of poetry] (Milan: Bompiani, 1968) and Gillo Dorfles' *Le oscillazioni del gusto* [Variations in taste] (Milan: Lerici, 1959).

CHAPTER FOUR

An essential study on the identification and description of the peculiarities of Renaissance perspective is Erwin Panofsky's *Die Perspektive als "symbolische Form"* [Perspective as symbolic form] (Leipzig, 1927). From Panofsky's enormous output, other works of interest here are, *The Life and Art of Albrecht Dürer* (Princeton, NJ: Princeton University Press, 1955), dedicated to one of the great founders of Renaissance perspective, and *Renaissance and Renascences in Western Art* (London: Paladin, 1970), where he returns to the notion of the uniqueness of the Renaissance, not to be confused with the less specific renascences that had previously occurred. Therefore, we have to conclude that the perspectives of Alberti and his contemporaries have new and unmistakable features. Naturally, we cannot avoid referring directly to the texts of the *homines novi* [new men] of the 1400s: Leon Battista Alberti's *Della pittura* [On painting], edited by L. Mallé (Florence: Sansoni, 1950); Piero della Francesca's *De prospectiva pingendi* [On perspective in painting], edited by Giusta Nicco Fasola (Florence: Sansoni, 1974); and Leonardo da Vinci's *Il trattato della pittura* [Treatise on painting], edited by Adachiara Zevi (Milan: Semir, 1982).

More recent works on the history and theory of perspective are Luciano Bellosi's "La rappresentazione dello spazio" [The representation of space] in *Storia dell'arte italiana* [History of Italian art] (Turin: Einaudi, 1982, 9 vols.) and Hubert Damisch's *L'origine de la perspective* [The origin of perspective] (Paris: Flammarion, 1987). The bold thesis asserting perspective as homologous to Gutenberg's invention is attributable, as stated above, to McLuhan's most famous works, in particular, *Through the Vanishing Point: Space in Poetry and Painting* (New York: Harper and Row, 1968). For an exhaustive and meticulous investigation of the origin and spread of the typographical revolution, we have Elizabeth Eisenstein's massive work, *The Printing Press as an Agent of Change* (Cambridge, UK: Cambridge University Press, 1979, 2 vols.), especially the chapter titled "The Unacknowledged Revolution," which essentially confirms McLuhan's views, although Eisenstein proceeds much more cautiously.

Giorgio Vasari's *Le vite de' più eccellenti pittori, scultori, e architettori* [Lives of the most excellent Italian painters, sculptors, and architects] can be consulted in Rosanna Bettarini's edition with a comment by Paola Barocchi (Florence: Sansoni, 1966, 9 vols.), which contains both the 1550 and the 1568 edition that includes the *Proems*). Barocchi also deserves mention for her work as editor of *Trattati d'arte del Cinquecento tra Manierismo e Controriforma* [Fifteenth-century treatises on art from Mannerism to the Counter-Reformation] (Milan and Naples: Ricciardi, 1971–72, 3 vols.). For an overview, I refer the reader to my contribution in *Letteratura italiana storia e testi* [Italian literature: History and texts] (Rome-Bari: Laterza, 1973, vol. iv, pt. 2). The piece also proposes a connection with the theories of the day pertaining to literary criticism, for which it is also useful to mention my *Poetica e retorica* [Poetics and rhetoric] (Milan: Mursia, 1982). Returning to Vasari, we have the monumental *Atti del Congresso Internazionale nel IV centenario della morte* [Proceedings from the international congress on the 4th centenary of Vasari's death] (Florence: Sansoni, 1976). On Italian art of the 1400s and the alternation of the spread of perspective and resistance to it offered by the pre-modern spirit of late Gothic, see the most popular and authoritative handbooks, including Giulio Carlo Argan's *Storia dell'arte italiana* [History of Italian art] (Florence: Sansoni, 1968, 3 vols., plus a fourth volume dedicated to modern art] and

Storia dell'arte italiana [History of Italian art], edited by C. Bertelli, G. Briganti, and A. Giuliano (Milan: Electa/Mondadori, 1986, 4 vols.). Also useful is *Storia dell'arte italiana* [History of Italian art] (Turin: Einaudi, 1982, 9 vols.), especially pt. 2, *Dal Medioevo al Novecento* [From the Middle Ages to the twentieth century]. See in particular Federico Zeri's *Rinascimento e Pseudo-Rinascimento* [Renaissance and pseudo-Renaissance, vol. 5] (1983) where a distinction is made between those who adopt perspective and those who resist it. Also fundamental is vol. 6, with articles by G. Romano, A. Pinelli, L. Spezzaferro, N. Spinosa, M. Mariani, L. Salerno, and A. Griseri, who review in depth and with an ample bibliography the stages of development presented in my own study, from the crisis of the Renaissance, Mannerism, Classicism, the Baroque, and seventeenth-century naturalism, up to the crisis of this last movement and the advent of eighteenth-century stylistic issues. Dealing more directly with the Renaissance-Mannerism phenomenon, see the seminal works of: Frederick Antal, *La pittura italiana tra Classicismo e Manierismo* [Italian painting from Classicism to Mannerism] (Turin: Editori Riuniti, 1977); Anthony Blunt, *Artistic Theory in Italy*, 1450–1600 (Oxford: Clarendon Press, 1940); Giuliano Briganti, *La maniera italiana* [The Italian *maniera*] (Florence: Sansoni, 1985); John Shearman, *Funzione e illusione: Raffaello, Pontormo, Correggio* [Function and illusion: Raphael, Pontormo, Correggio] (Milan: Il Saggiatore, 1983); and *The Renaissance and Mannerism: Acts of the Twentieth International Congress on the History of Art* (Princeton: Princeton University Press, 1963, 2 vols.); James V. Mirollo, *Mannerism and Renaissance Poetry* (New Haven: Yale University Press, 1984); and Antonio Pinelli, *La bella maniera* [The beautiful style] (Turin: Einaudi, 1993). More daring, because of their tendency to exaggerate, are the theories of Arnold Hauser, *Mannerism: The Crisis of the Renaissance and the Origin of Modern Art* (New York: Alfred A. Knopf, 1965) and especially Gustav René Hocke, *Die Welt als Labyrinth* [The world as labyrinth] (1957), where the subtitle indicates eloquently the mania for annexation: *Manier und Manie in der europäischen Kunst von 1520 bis 1650 und der Gegenwart*) [*Maniera* and mania in European art, from 1520 to 1650 and today], from which it is clear that the author extends the category of Mannerism to include the Counter-Reformation and the Baroque. Instead, I am in favour of respecting these traditional categories and their chronology and attempts to provide them with a more rigorous foundation. For the first movement, see Federico Zeri's *Pittura e Controriforma* [Painting and the Counter-Reformation] (Turin: Einaudi, 1957) and more generally my entry devoted to this period in *Letteratura italiana storia e testi* [Italian literature: History and texts] (Rome-Bari: Laterza, 1973, vol. iv, pt. 2). See also my *Maniera moderna e manierismo* [Modern *maniera* and Mannerism] (Milan: Feltrinelli, 2004).

There is an immense amount of literature on the topic of the Baroque. We can distinguish between works that approach the phenomenon primarily from the standpoint of literature and those that approach it from the standpoint of art. Those in the first group are inclined to emphasize the "conceptual" aspects of that stylistic extravagance, not without expressing some fondness for the theoreticians of Mannerism and their attempt to remove the barriers between the two movements. See, for example, Eugenio d'Ors, *Du Baroque* [On the Baroque] (Paris: Gallimard, 1968), as well as Luciano Anceschi's "Le poetiche del Barocco letterario in Europa" [The poetics of the literary Baroque in Europe] in *Momenti e problemi di storia dell'estetica* [Moments and problems in the history of aesthetics] (Milan: Marzorati, 1959, vol. i). Those in the second group are inclined instead to underscore the "persuasion" and eloquence of the Baroque phenomenon. See the writings of Giulio Carlo Argan, starting with the one included in *Retorica e Barocco* [Rhetoric and the Baroque], the proceedings from a homonymous conference held at the Fondazione G. Cini in Venice (Rome: Bocca, 1955) as well as *Immagine e persuasione. Saggi sul Barocco* [Image and persuasion: Essays on the Baroque], edited by Bruno Contardi (Milan: Feltrinelli, 1986). Here, Argan's approach intersects with the equally conciliatory and inclusive one taken by Cesare Gnudi in *L'ideale classico* [The classical ideal] (Bologna: Alfa, 1981). Despite its modest appearance, this volume can be considered as the theoretical nucleus and guide for the long and beautiful series *Biennali d'arte antica* [Biennial exhibits of ancient art], largely attributable to Gnudi and held in Bologna, with the occasional display in other cities of the Emilia-Romagna region. Because the Bologna school of the seventeenth century was naturally conciliatory and dialectical, the catalogues for these exhibits (all printed in Bologna by Alfa and subsequently Nuova Alfa) represent additional contributions to the phenomenology of interconnections between the dominant currents (Classicism, Naturalism, Baroque) and they reiterate the impossibility of making too sharp a distinction between these currents. They recommend, instead, that they be integrated in a unified image of the seventeenth century, even though that image is obviously vibrant and dynamic at its core. This corresponds very well to my own theory (presented in *Letteratura italiana storia e* testi as cited above) that is resumed in other forms outside the institutional context, owing to Andrea Emiliani's efforts to preserve the legacy: *Nell'età di Correggio e dei Carracci* [In the age of Correggio and the Carracci) (Bologna: 1986); *Guido Reni* (Florence: Giunti, 1988); *Giuseppe Maria Crespi* (Bologna: Nuova Alfa, 1990); and *Guercino* (1991).

CHAPTER FIVE

To deal with the late 1700s and early 1800s, still a largely ambiguous and confused area from the historiographic standpoint, there are many useful and informative texts, including the following: Robert Rosenblum and H.W. Janson, *19th-Century Art* (New York: Abrams, 1984); Hugh Honour, *Neo-classicism* (Harmondsworth: Penguin, 1977) and *Romanticism* (Boulder, Colorado: Westview Press, 1979); and Alfredo De Paz, *La rivoluzione romantica* [The romantic revolution] (Napoli: Liguori, 1984). This last work opens up to literary criticism and the history of ideas, a sector in which René Wellek's *A History of Modern Criticism 1750–1950, The Romantic Age* (New Haven: Yale University Press, 1955, 2 vols.) remains indispensable. Still within the literary field, see also Meyer Howard Abrams, *The Mirror and the Lamp: Romantic Theory and the Critical Tradition* (New York: Oxford University Press, 1953); Philippe Lacoue-Labarthe and Jean-Luc Nancy, *L'Absolu littéraire* [The literary absolute] (Paris: Seuil, 1978). There is certainly no shortage of valid studies on Italian art for the period in question, starting with Corrado Maltese's still solid *Storia dell'arte in Italia 1785–1945* [History of art in Italy 1785–1945] (Turin: Einaudi, 1960). See as well the articles by Anna Ottani and Sandra Pinto in the volume *Il Settecento e Ottocento* [The 1700s and 1800s] in *Storia dell'arte italiana* [The history of Italian art] (Turin: Einaudi, 1982, 9 vols.) in addition to the anthology *L'Ottocento* [The 1800s], edited by E. Castelnuovo (Milan: Electa, 1990).

Beyond these general studies, it is useful to examine more deeply the individual protagonists, as I have attempted to do in monographic university courses and the relevant notes, which were published as *L'alba del contemporaneo* [The dawn of the contemporary age] (Milan: Feltrinelli, 1996), as mentioned above. These courses were dedicated to Johann-Heinrich Füssli, Francisco Goya, Jacques-Louis David, William Blake, and William Turner, in that order. For a more detailed analysis of these authors see the monographs dedicated to them in the series *Classici dell'arte* (Milan: Rizzoli) for the first two figures and the last, while for David see Antoine Schnapper, *David, témoin de son temps* [David, witness of his time] (Fribourg: Office du Livre, 1980) and for Blake see Kathleen Raine's monograph (London: Thames and Hudson, 1970). In the case of Füssli, we have the excellent *Studi su Füssli* [Studies on Füssli] by Frederick Antal (Turin: Einaudi, 1971). The famous scholar also deserves mention for his essay *Hogarth e l'arte europea* [Hogarth and European art] (Turin: Einaudi, 1990), Hogarth being a fundamental figure in the institution of a "minor" current from which the modernist protest stems.

Not quite as valid is another of Antal's works, *Classicismo e Romanticismo* [Classicism and Romanticism] (Turin: Einaudi, 1975) in which the author insists on assigning a forward-looking role to the Géricault-Delacroix group instead of seeing their work as a return to the conventional space of modernity, as I suggest in the present study. Complementing the study of Füssli is Peter Tomory's *The Life and Art of Heinrich Füssli* (London: Thames and Hudson, 1972). For Goya, I refer the reader to two monumental works by José Gudiol, *Goya: biografia, estudio analitico y catalogo de sus pinturas* [Goya: biography, analytical study and catalogue of his paintings] (Barcelona, 1970) and Enrique Lafuente Ferrari, *Vie et oeuvre de Francisco Goya. L'oeuvre complet illustré: peintures, dessins, gravures* [Life and complete works of Francisco Goya: Etchings, aquatints, and lithographs] (Paris: Office du livre, 1970). Also useful and informative is Alfredo De Paz's *Goya* (Napoli: Liguori, 1990). A beautiful text that claims the figure of David was produced by one of his young pupils, Étienne-Jean Delécluze, *Jacques-Louis David. Son école et son temps* [Jacques-Louis David: His school and times] (Paris: Editions Macula, 1855). For these figures, see also Werner Hofmann, *Dal Neoclassicismo al Romanticismo* [From Neoclassicism to Romanticism] (Milan: Rizzoli, 1995).

In order to understand Blake, it is not enough to study his illuminations; we also need to take into account his literary production as well as his essays, especially in the light of the broad perspective within which he is considered in the present volume. See *Opere* [Works], edited by Roberto Sanesi (Milan: Guanda, 1984) as well as *Libri profetici* [Prophetic books] by the same editor (Milan: SE, 1987). Among the critical writings on Blake, see Sergio Givone, *William Blake. Arte e religione* [William Blake: Art and religion] and my article, "William Blake alle origini dell'età tecnotronica" [William Blake and the origin of the technotronic age], in *Rivista di estetica*, no. 32 (1989), in which I develop the positions outlined here. Also worth mentioning is a useful publication that has become almost canonical and has encouraged scholars to focus on all these great figures that straddle the 1700s and 1800s. I am referring to the annual reviews edited by Corrado Gizzi on behalf of the Casa di Dante of Abruzzo, in Torre de' Passeri (Pescara), designed to present the great illuminators of the *Divine Comedy*. Blake, Füssli, and Flaxman are discussed in symposia held in 1983, 1985, and 1986 respectively, with accompanying catalogues (Milan: Mazzotta). I collaborated in this project, providing contributions to the analysis of the art of these figures and their shared antimodern attitude. Also important are the issues dedicated to other illustrators of Dante's work, including, Joseph Anton Koch, who is associated with the Nazarenes (1988) and Dante Gabriel Rossetti, leader of the Pre-Raphaelites (1984). A later issue dedicated to Botticelli (Milan: Electa, 1990) can also be considered relevant to our area of inquiry in that Botticelli was a great "Pre-Raphaelite" in the literal sense, so much so that he fits in with the Nazarene and Pre-Raphaelite tendencies of the nineteenth century even though they are separated by centuries. On the last two phenomena, see the catalogues for the exhibits held in the Galleria Nazionale d'Arte Moderna in Rome, edited by G. Piantoni (Rome: De Luca, 1981) and in the Tate Gallery, London (London: Penguin Books, 1984), respectively. On the Pre-Raphaelites, see also my study (Milan: Fratelli Fabbri, 1967) as well as Maria Teresa Benedetti's, *Dante Gabriel Rossetti* (Florence: Sansoni, 1984) and the *Burne-Jones* catalogue for the exhibit held at the Galleria Nazionale d'Arte Moderna in Rome, edited by M.T. Benedetti and G. Piantoni (Milan: Mazzotta, 1986).

Leaving aside this group of opponents of modernity, who nonetheless appear as forerunners of the contemporary age, in order to deal with those who, instead, lead the innovative Romantic ferment toward modernity, preparing the way for its final phase, i.e., Realism and Naturalism, an important role is played by Francesco Hayez in Italian art. For this artist, see Fernando Mazzocca, *Invito a Francesco Hayez* [Introduction to Francesco Hayez] (Milan: Rusconi, 1982), in addition to the catalogue for the monographic exhibit edited by Cristina Gozzoli (Milan: Electa, 1983), see Andrea del Guercio, *La pittura dell'Ottocento* [Nineteenth-century painting] (Turin: UTET, 1982) and the catalogue for the exhibit *Il secondo Ottocento italiano* [Italian art in the second half of the nineteenth century] (Milan: Mazzotta,

1988), edited by myself, where I underscore the transition from Romanticism to Realism. See also the subsequent exhibit, dedicated to the preceding period, *Il primo Ottocento italiano* [Italian art in the first half of the nineteenth century] (Milan: Mazzotta, 1992), which deals with other Italian protagonists, apart from Canova (Appiani, Giani, Landi, Minardi, Pinelli, etc.). We should also be aware of another exhibit, which I edited, *Impressionismo italiano* [Italian Impressionism] (Milan: Mazzotta, 2002), and, finally, my detailed overview in *Storia dell'arte contemporanea in Italia: Da Canova alle ultime tendenze* [The history of contemporary art in Italy: From Canova to the latest trends] (Turin: Bollati Boringhieri, 2007), which expands the meaning of the term to include the 1900s.

With respect to contemporary art, understood in the restricted sense suggested here (which is to say, from Cézanne to today), I limit myself to identifying the main phases of that art, paying particular attention to its connections with the general perspectives of culturology. Therefore, I do not provide an exhaustive bibliography. Instead, I refer the reader to my *L'arte contemporanea* (Milan: Feltrinelli, 2005), where I have collected systematically more specific articles or more widely circulated articles produced before and after the first edition of the book. See, for example, *Il Simbolismo* [Symbolism], vol. 2 of *L'arte moderna* [Modern Art] (Milan: Fratelli Fabbri, 1966), which is only a part of the larger work in 14 volumes. See also *Informale oggetto comportamento* [Informal object behaviour] (Milan: Feltrinelli, 2006].

CHAPTER SIX

The most up-to-date study on Wölfflin is Gian Luca Tusini's *Il fronte della forma* [The front of form] (Bologna: BUP, 2005), which contains useful comparisons with the critical theory of Alois Riegl, Erwin Panofsky, and Wilhelm Worringer. A few references might be useful for those who wish to examine further the great "philosophical operators," contemporaries of Wölfflin with whom they are homologous: Edmund Husserl's early work, *Logische Untersuchungen* [Logical investigations] (Halle a. d. S.: Max Niemeyer, 1913); the classic *Ideas: General Introduction to Pure Phenomena* (New York: Macmillan, 1931); *Erfahrung und Urteil* [Experience and judgment] (Prague: Academia Verlagsbuchhandlung, 1939); and *Die Krisis der europäischen Wissenschaften und die transzendentale Phänomenologie* [Crisis of European sciences and transcendental phenomenology] (Belgrade: Philosophia, 1936); the crisis in the title refers to the loss of grounding and the fragmentation of the sciences into a forest of specialized disciplines not coordinated with one another. Although it appears, at first glance, that he has a different view on the biological basis of experience, John Dewey is sensitive to wholes and general ideas, as we discover not only in *Art as Experience* (New York: Minton, Balch, 1934) but also in works on logic and epistemology, such as *Logic: The Theory of Inquiry* (New York: Henry Holt and Company, 1938) and *The Quest for Certainty: A Study of the Relation of Knowledge and Action* (New York: Minton, Balch, 1929), written in collaboration with Arthur Fisher Bentley. Two observations can be drawn from this, even if there is some ambiguity involved. One is that we have confirmation of the quest for primary "data," as announced by William Blake (see the preceding chapter), who struggled himself against the first manifestation of elementariness as it appeared in "modern" thought, especially Empiricism and Sensism, right up to the Enlightenment. We have seen, however, that Blake's effort to move toward "holism," with its aim of preserving the unity and compactness of perceived ideas, was unsuccessful. Soon, a new elementariness would take hold, one corresponding to the positivist period. From this stems the reaction of the late-century "holists," namely, Wölfflin and his followers who, however, had to contend with yet another revival of elementariness in the analytical and neo-positivist aspect of twentieth-century philosophy. These alternating phases recur even after the Second World War, oscillating among *Husserl Renaissance*, the main exponents of which are the members of Parisian phenomenological school (Jean-Paul Sartre, Maurice Merleau-Ponty, Mikel Dufrenne), the school of Enzo Paci, the school of Luciano Anceschi, and so-called critical phenomenology. On these issues, see my *Per un'estetica mondana* [Toward an earthly aesthetics] (Bologna: Il Mulino, 1964). The 1960s, however, are dominated by yet another flowering of analytics, whose future is entrusted to semiotics this time. A key publication in this period is Umberto Eco's *Trattato di semiotica generale* [A theory of semiotics] (Milan: Bompiani, 1975). The author was subsequently inspired to adopt more flexible criteria, as we see in *I limiti dell'interpretazione* [The limits of interpretation] (Milan: Bompiani, 1975), which opens up the opposite front, that is, against theories that are too liberal and concede to the reader unlimited interpretive abilities. In the meantime, the "school of absence" came into existence and prospered, represented primarily by French philosopher Jacques Derrida, with whose ideas Eco had come to terms in his *La struttura assente* [The absent structure] (Milan: Bompiani, 1968). See also my *Tra presenza e assenza* [Between presence and absence] (Milan: Bompiani, 1979].

The second observation regards the root of the methodology of Wölfflin, Husserl, Dewey, Bergson, etc. On one side, we can confirm the presence of "whole," compact ideals; on the other side, these same ideals could not be allowed to be static or inert; their ability to change needed to be assured. From this emerges the need, shared by all these thinkers, to imagine the ability not only to construct ideas, but also to deconstruct them. In other words, general ideas are vectorial and not scalar (inert, absolute, ahistorical). Out of this comes the system of polar opposites, for which Wölfflin's methodology is justifiably famous, but one that is confirmed some time later by Luciano Anceschi's approach, starting with his early work *Autonomia ed eteronomia* [Autonomy and heteronomy] (Florence: Vallecchi, 1936). There is no room, therefore, for a theory of art that is "closed" or fixed within a stable paradigm; instead, we should learn from his *Progetto di una sistematica dell'arte* [Project for a system of art] (Milan: Mursia, 1960), which allows the free development of the various processes whereby art is instituted and de-instituted within the framework of cultural forms. See *Le istituzioni della poesia* [The institutions of poetry] (Milan: Bompiani, 1968). This freedom is assured by the "descriptive" instruments that rhetoric provides, rather than by the "discrete" analytical tools of semiotics. As for the first, see chapter 2 of the present study. Rhetorical logic proves to be better suited than analytical logic to meet the demands of the humanistic disciplines. This was known in a long tradition established by Aristotle and continued by Husserl who, in his call to safeguard "ideas," also warned that these ideas should be selected so as to be commensurate with the field of experience to which they are applied. In other words, rigour is not necessarily geometric, and geometry is not based solely on regular figures. In this regard, see yet another fundamental work by Husserl, which I recommend in the French translation *L'Origine de la géométrie* [The origin of geometry] (Paris: PUF, 1962), with an introduction by Jacques Derrida, who derived from this work his own thinking on "différence," aimed at theorizing unlimited interpretive freedom. For a reading of Bergson that is in line with what is proposed here, see my *Bergson, il filosofo del software* [Bergson, the philosopher of software] (Milan: Cortina, 2005).

Index

Abrams, Meyer Howard and *The Mirror and the Lamp: Romantic Theory and the Critical Tradition*, 168

Abstract Expressionism, 130, 145

Agazzi, E. and C. Cellucci and *Aspetti storici, filosofici, e matematici della logica moderna e le sue applicazioni* [Historical, philosophical, and mathematical aspects of modern logic and its applications], 161

Alberti, Leon Battista, 7, 61–3, 67–8, 70, 92, 150, 164; *Della pittura* [On painting], 164

Anceschi, Luciano, 74, 150, 164, 167, 172–3; *Autonomia ed eteronomia dell'arte* [Autonomy and heteronomy of art], 150, 173; *Le istituzioni della poesia* [The institutions of poetry], 164; *Progetto di una sistematica dell'arte* [Project for a system of art], 173

Annales School, 160

Anquetin, Louis, 52, 121

Antal, Frederick, 166, 169; *Classicism and Romanticism*, 169; *Hogarth and European Art*, 169; *Italian Painting between Classicism and Mannerism*, 166; *Studies on Füssli*, 169

Anti-Form Art, 134

Antoni, Carlo, 147

Antoninus Pius, 100

Appiani, Andrea, 100, 171

Argan, Giulio Carlo, 74, 163, 165, 167; *Storia dell'arte italiana* [History of Italian art], 165

Aristotle, 31–3, 35, 59, 161, 173; *Organon*, 161; *Poetics*, 59

Armstrong Richards, Ivor and *The Philosophy of Rhetoric*, 162

Art Déco, 57

Art Nouveau, 51

Arte Povera, 134

Bacon, Francis, 75

Balzac, Honoré de, 111–2

Barilli, Renato, 83, 101, 118, 163, 171; *Bergson, il filosofo del software* [Bergson, the software philosopher], 172; *Comicità in Kafka*

[The comical in Kafka], 162; *Corso di retorica* [Course on rhetoric], 32; *Il ciclo del postmoderno* [The postmodern cycle], 77, 160; *L'alba del contemporaneo* [The dawn of the contemporary age], 84, 94, 102, 168; *L'enigma Pirandello* [The enigma of Pirandello], 162; *Maniera moderna e manierismo* [Modern maniera and Mannerism], 71, 167; *Per un'estetica mondana* [Toward an earthly aesthetics], 41, 163, 172 *Poetica e retorica* [Poetics and rhetoric], 33, 40, 71, 74, 165; *Retorica* [Rhetoric], 3, 162 *Storia dell'arte contemporanea in Italia da Canova alle ultime tendenze* [Contemporary art in Italy from Canova to recent trends], 83, 101, 118, 163, 171; *Tra presenza e assenza* [Between presence and absence], 173
barocchetto, 76, 95, 99
Barocchi, Paola, 68, 165; *Trattati d'arte del Cinquecento fra Manierismo e Controriforma* [Sixteenth-century art treatises between Mannerism and the Counter-Reformation], 165
Baroque, 51, 74–6, 89–91, 95–6, 99
Bassano, Jacopo, 71
Bauhaus, 130
Beato Angelico, 67
Behavioural Art, 134
Bell, Daniel, 160
The Coming of the Post-Industrial Society, 160
Bellini, Giovanni (Giambellino), 67
Benedetti, Maria Teresa and *Dante Gabriel Rossetti*, 170
Bentley, A.F., 172
Bergson, Henri, 43, 87, 89, 103, 136, 139, 162, 173; *élan vital*, 89, 103
Bernard, Émile, 62, 95, 121
Bernini, Gian Lorenzo, 74–5, 146–7, 149
Bertelli, Carlo, 166
Bettarini, Rosanna, 165
Biennale d'arte antica, 72, 76, 167
biomorphism, 128
Blake, William, 8, 84–91, 93–4, 96–8, 100–1, 104, 110–1; prophetic books, 84–5, 169; *Proverbs of Hell*, 88; *Songs of Experience*, 84; *Songs of Innocence*, 84; *The Marriage of Heaven and Hell*, 85–88
Blanche, R. and *The History of Logic from Aristotle to Russell*, 161
Blunt, Anthony and *Artistic Theory in Italy, 1450–1600*, 166
Boccioni, Umberto, 128, 154
Böcklin, Arnold, 114
Body Art, 134
Bohr, Niels, 34
Bonheur, Rosa, 114
Borromini, Francesco, 74
Botticelli, Sandro, 67, 70, 103, 107, 146–7, 151, 170
Bouguereau, William-Adolphe, 114
Braque, Georges, 52, 124–5, 128
Braudel, Ferdinand and *Capitalism and Material Life*, 160; *Between Two Ages: America's Role in the Technetronic Age*, 160
Breton, Jules, 114
Briganti, Giuliano and *La maniera*

italiana [The Italian *maniera* or style], 166
Bronzino (Agnolo di Cosimo), 71
Brzezinski, Zbigniew, 160
Butor, Michel, 56

Cabianca, Vincenzo, 114
Calvinism, 95
camera obscura, 22, 63, 92, 150
Canova, Antonio, 100, 105, 171
Caravaggio, Michelangelo Merisi da, 73–4, 108, 149
Cardinal Gabriele Paleotti, 72–3
Carrà, Carlo, 129
Carracci (family), 72, 96, 109, 146, 149, 168
Carracci, Agostino, 72
Carracci, Annibale, 72–3, 75–6, 105, 108, 147
Carracci, Ludovico, 73
Carstens, Asmus Jacob, 100
Cassirer, Ernst, 5, 26, 41, 161
Castagno, Andrea del, 67
Castelnuovo, Enrico, 168
L'Ottocento [The 1800s], 168
Castelvetro, Ludovico, 71
Cervantes, Miguel de and *Don Quixote*, 118
Cézanne, Paul, 64, 68, 70, 82–4, 101, 118–21, 125, 154, 171
Chirico, Giorgio de, 129
Cicero, 33
Cimabue, 93, 107
Classicism, 74–5, 100, 104–6, 109–10, 143, 149, 166–7, 169
Cloisonnism, 95, 121
Columbus, Christopher, 7
Conceptual Art, 134
concettismo, 74–5
Concrete Art, 57
Congress of Vienna, 105
Constructivism, 127, 130
Contardi, Bruno and *Immagine e persuasione: Saggi sul Barocco* [Image and persuasion: Essays on the Baroque], 167
Correggio, Antonio Allegri da, 89, 91, 96, 108, 148, 166, 168
Cortona, Pietro Berrettini da, 74, 108, 149
Counter-Reformation, 72–3, 148, 165–7
Courbet, Gustave, 114
Crespi, Giuseppe Maria, 22, 76, 168
Creti, Donato, 76
Crivelli, Carlo, 67
Croce, Benedetto, 137–8, 147, 150
Cubism, 30, 52, 123, 125, 127, 134, 145

Dada, 54–7, 127, 131
Damisch, Hubert and *L'Origine de la perspective* [The origin of perspective], 165
D'Annunzio, Gabriele, 53
Dante, 94, 170; *Divine Comedy*, 170
David, Jacques-Louis, 94, 99–100, 104–5, 107, 109–10, 144, 168–9; *The Oath of the Horatii*, 100
Decadentism, 51
Degas, Edgar, 113, 121
Delacroix, Eugène, 108–10, 112, 169
Delaunay, Robert, 128
Delécluze, Étienne-Jean and *Jacques-Louis David: Son école et son temps* [Jacques-Louis David : His school and his times], 169
Derrida, Jacques, 173

Descartes, René, 75
Dewey, John, 136, 162, 172–3; *Art as Experience*, 162, 172; *Logic: The Theory of Inquiry*, 172; *The Quest for Certainty*, 162, 172
Dine, Jim, 132
Divisionism, 52, 121, 123
Dorfles, Gillo and *Le oscillazioni del gusto* [Fluctuations in Taste], 164
D'Ors, Eugenio and *Du Baroque* [On the Baroque], 167
Duchamp, Marcel, 127, 129, 131
Dufrenne, Mikel, 172
Dürer, Albrecht, 146, 164

Eco, Umberto, 25, 58, 164, 172–3; *I limiti dell'interpretazione* [The limits of interpretation], 172; *La struttura assente* [The absent structure], 173; *L'opera aperta* [The open work], 164; *Trattato di semiotica generale* [A theory of semiotics], 172
Eddas, 96
Einstein, Albert, 38, 81
Eisenstein, Elizabeth and *The Printing Press as an Agent of Change*, 165
electromagnetism, 8, 38, 80–2, 87–8, 91, 96, 102
Emiliani, Andrea and *Nell'età di Correggio e dei Carracci* [The age of Correggio and the Carracci family], 168
Empiricism, 27, 36, 86, 96, 172
Enlightenment, 86, 96, 172
Ensor, James, 52, 122, 128
Ernst, Max, 130
Euclidian geometry, 63–5
Evergreen, 121
Expressionism, 128, 130, 145

Fattori, Giovanni, 114
Fauves, 123, 128
Febvre, Lucien and Henri-Jean Martin and *The Coming of the Book*, 60
Feuerbach, Anselm, 114
Fiedler, Konrad, 163
Flaubert, Gustave and *The Temptation of Saint Anthony*, 118
Flaxman, John, 94–5, 100, 110, 121, 170
Fluxus, 134
Focillon, Henri and *La vie des formes* [The life of forms], 164
Fra Bartolomeo, 70, 146
Fragonard, Jean-Honoré, 99
French Revolution, 7, 78–9, 105
Freud, Sigmund, 43, 48, 87–9, 97–8, 162–3; Ego, 58; Id, 58; Libido, 58; Super-ego, 58
Friedrich, David Caspar, 101, 103
Füssli, Johann Heinrich (Henry Fuseli), 8, 94–101, 104, 168–70
Futurism, 55, 57, 128

Gainsborough, Thomas, 99
Galileo Galilei, 58, 75
Galvani, Luigi, 7, 84
Gauguin, Paul, 52–3, 95, 98, 105, 121–3, 132
Gaulli, Giovanni Battista, 74
Géricault, Théodore, 108–10, 112, 169; *The Raft of the Medusa*, 109
Ghirlandaio, Domenico, 67, 146, 151

Giani, Felice, 101, 171
Giorgione (Giorgio Barbarelli), 70, 91, 108
Giotto, 69, 93, 107
Giuliano, Antonio, 116
Givone, Sergio and *William Blake: Arte e religione* [William Blake: Art and religion], 169
Gizzi, Corrado, 94, 170
Gleizes, Albert, 128
Goldmann, Lucien, 6, 19, 159, 161; homology, 6, 19, 22, 27, 60, 66, 85, 130, 134, 150, 152, 159, 164; *Introduction à la philosophie de Kant* [Introduction to the philosophy of Kant], 159; *Pour une sociologie du roman* [Towards a sociology of the novel], 159; *Recherches dialectiques* [Dialectical investigations], 159, 161
Gnudi, Cesare, 72, 167; *L'ideale classico* [The classic ideal], 167
Goethe, Wolfgang, 86–7; *Elective Affinities*, 86
Gothic, 98, 115, 123, 165
Goya, José Francisco, 94, 97–100, 121, 168–9
Gozzoli, Cristina, 170
Gracián, Baltasar, 74–5
Griseri, Andreina, 166
Gropius, Walter, 130
Gudiol, José and *Goya: Biografia, estudio analitico y catalogo de sus pinturas* [Goya: Biography and catalogue of his paintings], 169
Guercino, 72, 76, 168
Guercio, Andrea del and *La pittura dell'Ottocento* [Nineteenth-century painting], 170
Gurvitch, Georges and *Dialectique et sociologie* [Dialectics and sociology] 161
Gutenberg, Johannes, 7, 22–3, 27, 29, 33, 36, 60–1, 64, 66, 69, 71–2, 79, 83, 87, 150, 165

Harris, Marvin, 3, 6, 13, 15, 157; *America Now: The Anthropology of a Changing Culture*, 157; *Cannibals and Kings: The Origin of Cultures*, 15, 157; *Cultural Materialism: The Struggle for a Science of Culture*, 3, 157; *The Rise of Anthropological Theory: A History of Theories of Culture*, 13, 157
Hassan, Ihab and *The Dismemberment of Orpheus: Toward a Postmodern Literature*, 160
Hatoum, Mona, 155
Hauser, Arnold and *Mannerism: The Crisis of the Renaissance and the Origin of Modern Art*, 166
Hayez, Francesco, 110, 112–4, 170
Hegel, Georg Wilhelm Friedrich, 116–7, 137
Heisenberg, Werner, 39, 58; Uncertainty Principle, 58
Hildebrand, Adolf von, 163
Hocke, Gustave René and *The World as Labyrinth*, 166
Hodler, Ferdinand, 122
Hofmann, W. and *From Neoclassicism to Romanticism*, 169
Homer, 95
Honour, Hugh, 94, 168; *Neoclassicism*, 94, 168
Humanism, 33, 38, 66, 70
Husserl, Edmund, 12, 136, 138,

171–3; *Crisis of European Sciences and Transcendental Phenomenology*, 171; *Experience and Judgment*, 171; *Ideas: General Introduction to Pure Phenomena*, 171; *Logical Investigations*, 171; *Husserl Renaissance*, 172
Hutcheon, Linda and *A Poetics of Postmodernism*, 160
Hunt, William, 115

Impressionism, 23, 119, 120–1, 145, 171
Industrial Revolution, 29–30, 65, 76, 79, 83
Informal Art, 50
Ingres, Jean-Auguste Dominique, 107–8, 144
Innis, Harold Adams and *The Bias of Communication*, 158

Janson, H.W. and *Nineteenth-Century Art*, 94, 168
Jencks, Charles and *The Language of Post-modern Architecture*, 160
Joyce, James, 55–6
Jugendstil, 51, 121

Kafka, Franz, 56, 162
Kahn, Gustave, 53
Klimt, Gustav, 122, 128
Kluckhohn, Clyde and Kroeber, Louis, 3, 13, 157; *Culture: A Critical Review of Concepts and Definitions*, 3, 13
Koch, Joseph Anton, 170
Kubler, George and *The Shape of Time: Remarks on the History of Things*, 164

La Voce, 55
Lacoue-Labarthe, Philippe and Jean-Luc Nancy and *L'Absolu littéraraire* [The literary absolute], 168
Lafuente Ferrari, Enrique and *Life and Complete Works of Francisco Goya: Etchings, Aquatints, and Lithographs*, 169
Land Art, 134
Lanfranco, Giovanni, 108
Langer, Susanne, 5, 26, 41, 161; *Feeling and Form: A Theory of Art*, 161; *Philosophy in a New Key*, 26, 41
Laplanche, Jean and Jean-Bertrand Pontalis and *Enciclopedia della psicoanalisi* [Encyclopedia of psychoanalysis], 163
Lega, Silvestro, 114
Léger, Fernand, 128
Leopardi, Giacomo, 86, 111–13
Lichtenstein, Roy, 132
Lippi, Filippo, 67
Lukács, Georg, 19
Luperini, Romano, 163
Il Novecento [The 1900s], 163
Lyotard, Jean-François and *La Condition postmoderne* [The postmodern condition], 160

Macchiaioli, 113–14
Maeterlinck, Maurice, 53
Malevich, Kasimir, 127
Maltese, Corrado and *Storia dell'arte in Italia 1785–1945* [History of Italian art 1785–1945], 168
Manet, Edouard, 113, 118–19; *Déjeuner sur l'herbe* [The luncheon on the grass], 118; *Olympia*, 118

Mannerism, 71–2, 96, 101, 148, 165–7
Mantegna, Andrea, 67, 103
Manzoni, Alessandro, 110–12; *I promessi sposi* [The betrothed], 110
Marées, Hans von, 163
Mariani, Carlo Maria, 166
Marino, Giambattista, 75
Martin, Henri Jean, 60, 157; *Histoire et pouvoir de l'écrit* [The history and power of writing], 60, 157
Marxism, 4, 19, 117, 159
Masaccio, 67, 107
Matisse, Henri, 52
Maxwell, James Clerk, 27, 34, 38, 82, 87, 118; *Treatise on Electricity and Magnetism*, 34, 82
Mazzocca, Fernando and *Invito a Francesco Hayez* [Introduction to Francesco Hayez], 170
McLuhan, Corinne, 159
McLuhan, Eric and *Laws of Media: The New Science*, 159
McLuhan, Marshall, 4, 7–8, 21–3, 27, 29, 36, 38, 60–1, 63–6, 78–9, 137, 150–3, 158–9, 164–5; *The Gutenberg Galaxy: The Making of Typographic Man*, 21, 79, 154, 158; *The Mechanical Bride: Folklore of Industrial Man*, 158–9; *Through the Vanishing Point: Space in Poetry and Painting*, 63, 65; *Understanding Media: The Extensions of Man*, 4, 21, 158
mechanomorphism, 23, 30, 120, 127
Medici, Lorenzo de', 7, 79
Melandri, Enzo and *La linea e il circolo* [The line and the circle], 161
Merleau-Ponty, Maurice, 172
Messina, Antonello da, 67
Metaphysical Painting, 57, 129, 134
Metzinger, Jean, 128
Michelangelo, 29, 70, 90–1, 93–4, 96, 110
Millais, John Everett, 115
Millet, Jean-François, 114
Minardi, Tomaso, 171
Minimalism, 134
Mirollo, James V. and *Mannerism and Renaissance*, 166
Mitscherlich, Alexander and *Society without the Father*, 163
Modernism, 51
Molinaro, Marie, 159
Mondrian, Piet, 127, 154
Morandi, Giorgio, 129
Morelli, Domenico, 114–15; *Eleonora Duse*, 114; *The Iconoclasts*, 114; *Torquato Tasso*, 114
Mori, Mariko, 155
Morris, William, 69
Mukarovsky, Jan, 163; *Aesthetic Function, Norm, and Value as Social Facts*, 163; *Studies in Aesthetics*, 163
Munch, Edvard, 52, 122, 128

Nabis, 53, 105, 122
Napoleon, 100, 105
Naturalism, 20, 27, 37, 73–4, 88, 91, 96, 106, 109–10, 112–13, 115–6, 149, 166–7, 170
Nazarenes, 104–8, 115, 151, 170
Neo-avant-garde, 50, 55
Neoclassicism, 94, 99–100, 105, 107, 110, 144–5, 169
Neoplasticism, 127

Neshat, Shirin, 155
Newton, Isaac, 58, 82, 84–5, 87, 89, 97
Nibelung, 96
Nicco Fasola, Giusta, 165
Novissimi, 55

Olbrechts-Tyteca, Lucie, 162
Oldenburg, Claes, 132
Ong, Walter, 159, 162; *Interfaces of the Word: Studies in the Evolution of Consciousness and Culture*, 159; *Orality and Literacy: The Technologizing of the World*, 159; *The Presence of the Word: Some Prolegomena for Cultural and Religious History*, 159
Ottani, Anna, 168
Overbeck, Friedrich, 104–5, 107

Paci, Enzo, 172
Pacinotti, Antonio, 27, 80, 82, 117
Pallavicino, Sforza, 74
Panofsky, Erwin, 5–6, 20–1, 60, 63–4, 149–50, 152, 164, 171; *Perspective as "Symbolic Form,"* 164; *Renaissance and Renaissances in Western Art*, 5, 164; *The Life and Art of Albrecht Dürer*, 164
Parmigianino, 71, 148
Pascoli, Giovanni, 53
Paz, Alfredo de, 168–9; *Goya*, 169; *La rivoluzione romantica* [The romantic revolution], 168
Pellegrini, Matteo, 74
Perelman, Chaim and *Traité de l'argumentation: La nouvelle rhétorique* [The new rhetoric: A treatise on argumentation], 162
perspective, 5, 14, 20–3, 25, 33, 53, 60–1, 63–4, 66–70, 75–6, 81–2, 87, 97, 102, 118, 150–2, 158, 164–6, 169, 171
Perugino, Pietro, 67, 90, 108
photography, 8, 23, 62–3, 92–3, 98, 122, 150
phytomorphism, 121, 124
Piantoni, Gianna, 170
Picasso, Pablo, 52, 124–6, 128, 154
Piero della Francesca, 67, 107, 164; *De prospectiva pingendi* [On perspective for painting], 164
Pinelli, Antonio, 166, 171; *La bella maniera* [The beautiful style], 166
Pinturicchio (Bernardino di Betto), 67
Piombo, Sebastiano del, 70
Pissarro, Camille, 113, 119, 121
Pforr, Franz, 104
Planck, Max, 39
Plato, 66, 85
Platonism, 85, 137
Plotinus, 85
Pollaiuolo, Antonio del, 67
Pont-Aven School, 105, 122
Pontormo, Jacopo, 71, 148, 166
Pop Art, 50, 132
Positivism, 4, 66, 88, 117, 141–2
Postmodernism, 109, 113, 160
Poussin, Nicolas, 73–4, 99, 105, 108–9, 112, 149
Pre-Raphaelites, 69, 90, 100, 103, 105, 107, 115, 170
Preti, Giulio and *Retorica e logica* [Rhetoric and logic], 32, 38, 162
Previati, Gaetano, 53, 122, 128
Process Art, 134
Primitivism, 100, 106–7, 116

pure visibility, 163
Puvis de Chavannes, Pierre, 114

Raine, Kathleen and *William Blake*, 169
Raphael Sanzio, 29, 69–70, 90–1, 93, 99, 104–8, 110, 115, 129
Rationalism, 27, 36, 72, 86, 96
Reformation, 72, 78
Rella, Franco and *Forme e pensiero del moderno* [Forms and ideas of the modern], 160
Rembrandt van Rijn, 74, 89, 46
Reni, Guido, 72, 76, 108, 168
Renoir, Pierre-Auguste, 52, 113
Reynolds, Joshua, 89, 97, 99
Richards, Ivor Armstrong and *The Philosophy of Rhetoric*, 162
Riegl, Alois, 5, 163, 171
Robbe-Grillet, Alain, 56
Rococo, 76, 95, 99, 105, 109, 144
Romanesque, 115
Romano, Giulio (Pippi), 71, 163, 166
Romanticism, 39, 94, 102, 106, 110–12, 168–9, 171
Römer, 114
Rosenblum, Robert, 94, 168
Rosenquist, James, 132
Rossetti, Dante Gabriel, 104, 115, 170
Rubens, Peter Paul, 74, 89, 146
Ruysdael, Salomon van, 74

Salerno, Luigi, 166
Sanesi, Roberto, 169
Sarto, Andrea del, 70, 146
Sartre, Jean-Paul, 172
Saussure, Ferdinand de, 43, 136, 138–41, 143, 147; *Cours de linguistique générale* [Course on general linguistics], 43, 130
Schelling, Friedrich Wilhelm Joseph, 116
Schlegel, August, 116
Schlegel, Friedrich, 116
Schnapper, Antoine and *David, témoin de son temps* [David, witness to his times], 169
Scholz, Heinrich and *A Concise History of Logic*, 161
School of Dynamic Formalism, 164
Schopenhauer, Arthur, 87, 116; *The World as Will and Representation*, 87
Segantini, Giovanni, 52–3, 122
Seurat, Georges, 52–3, 98, 122–3, 132
Shearman, John, 166; *Function and Illusion: Raphael, Pontormo, Correggio*, 166; *The Renaissance and Mannerism. Acts of the Twentieth International Congress on the History of Art*, 166
Signorelli, Luca, 67, 70, 103, 107
Signorini, Telemaco, 113–14
Simon, Claude, 56, 157
simulacrum, 24, 62
Sisley, Alfred, 52, 113
Snow, C.P. and *The Two Cultures*, 38, 161
Spezzaferro, Luigi, 166
Spinosa, Nicola, 166
Stendhal (Marie-Henri Beyle), 111
Suprematism, 127
Surrealism, 130
Symbolism, 51, 53, 98, 112, 121, 128, 132, 163, 171

Synthetism, 52, 121, 123

Tabaimo, 155
Tachisme, 51, 130, 145
Tanguy, Yves, 130
Tatlin, Vladimir, 127
techne, 14, 28
technomorphism, 20, 80, 124
terribilità, 106
Tesauro, Emanuele, 74–5
Tibaldi, Pellegrino, 71
Tiepolo, Giambattista, 99, 110
Tintoretto (Jacopo Robusti), 71
Titian (Tiziano Vecellio), 70, 89, 91, 108, 110, 146
Toffler, Alvin and *The Third Wave*, 160
Tomory, Peter and *The Life and Art of Heinrich Füssli*, 160
Trajan's column, 100
trobar clos, 71
Tura, Cosimo, 67
Turner, Joseph Mallord William, 101–3, 109, 168
Tusini, Gian Luca and *Il fronte della forma* [The front of form], 61

Uccello, Paolo, 67, 107

Valente, T., 159
Valori Plastici, 57
Vasari, Giorgio, 61, 68, 70, 90, 146, 151, 165; *Le vite de' più eccellenti pittori, scultori e architettori* [The lives of the most excellent Italian painters, sculptors, and architects], 68
Veneziano, Domenico, 67
Venturi, Lionello and *Storia della critica d'arte* [History of art criticism], 163
Ver sacrum, 121
Veronese, Paolo, 110
Verrocchio, Andrea del, 67, 146
Vico, Giambattista, 117
Vinci, Leonardo da, 68–70, 94, 108, 120, 122, 146, 151, 165; *Il trattato della pittura* [Treatise on painting], 165
Volta, Alessandro, 7, 84

Warhol, Andy, 132
Wellek, René and *A History of Modern Criticism, 1750–1950*, 168
Wesselmann, Tom, 132
White, Leslie, 3, 6, 26, 45, 158, 162; *The Science of Culture: A Study of Man and Civilization*, 3, 26, 158
Wilde, Oscar, 53
Wölfflin, Heinrich, 5, 57–8, 68–70, 73, 137–52, 164, 171–3; closed forms, 57–8, 69–70, 147; open forms, 57–8, 69–70, 145, 147–8
Woolf, Virginia, 56